HOLISM AND THE UNDERSTANDING OF SCIENCE

'This book addresses issues which are central in the philosophy of science, exploring a large and relevant literature. It should be of broad interest in the philosophy of science community.'

-Professor Peter Lipton, Department of History and Philosophy of Science,
University of Cambridge, UK

How can the complexities of understanding science be dealt with as a whole? Is philosophical realism still a defensible philosophical position? Exploring such fundamental questions, this book claims that science ought to be understood in terms of universal practices and that such an understanding supports an attractive version of scientific realism. Holism is attracting renewed scholarly attention but is still loosely used in a range of different contexts, from semantics to medicine. This book presents a detailed philosophical analysis of holism, concentrating on two complementary aspects of holism - cognitive and social - to investigate its relevance to science studies. Bridging the gap between analytical, historical and sociological accounts of science, Caruana draws together results from recent research by Davidson, Dummett, Quine, Wright and others, on Wittgenstein's later philosophy.

Demonstrating that holism, both cognitive and social, is not only essential for a full understanding of science but also compatible with a particular version of scientific realism, this book presents important new perspectives for the philosophers of science and scholars of the history of science in particular.

Louis Caruana is Assistant Lecturer in Philosophy of Science and Nature at the Gregorian University, Rome, Italy.

T0251123

ASHGATE NEW CRITICAL THINKING IN PHILOSOPHY

The *Ashgate New Critical Thinking in Philosophy* series aims to bring high quality research monograph publishing back into focus for authors, the international library market, and student, academic and research readers. Headed by an international editorial advisory board of acclaimed scholars from across the philosophical spectrum, this new monograph series presents cutting-edge research from established as well as exciting new authors in the field; spans the breadth of philosophy and related disciplinary and interdisciplinary perspectives; and takes contemporary philosophical research into new directions and debate.

Series Editorial Board:

Holism and the Understanding of Science

Integrating the analytical, historical and sociological

LOUIS CARUANA

Routledge
Taylor & Francis Group

LONDON AND NEW YORK

First published 2000 by Ashgate Publishing

Reissued 2018 by Routledge
2 Park Square, Milton Park, Abingdon, Oxon OX14 4RN
711 Third Avenue, New York, NY 10017, USA

Routledge is an imprint of the Taylor & Francis Group, an informa business

Publisher's Note
The publisher has gone to great lengths to ensure the quality of this reprint but points out that some imperfections in the original copies may be apparent.

Disclaimer
The publisher has made every effort to trace copyright holders and welcomes correspondence from those they have been unable to contact.

A Library of Congress record exists under LC control number: 00133520

ISBN 13: 978-1-138-73493-7 (hbk)
ISBN 13: 978-1-138-73491-3 (pbk)
ISBN 13: 978-1-315-18680-1 (ebk)

Contents

Preface

This book is about ways of dealing with the complexity one faces when trying to understand science as a whole. In general, when giving an account of something, we tend to simplify our work by identifying the diverse constituent aspects which seem to us to call for explanation. This strategy, however, may result in the systematic neglect of some important aspects which are only perceptible when the thing is taken as a whole. As regards science, various attempts have been made to understand it through a philosophical analysis of theoretical terms, or of equations, or theories. Such understanding in terms of apparently isolated units, however, does not do justice to the fact that we can fully understand these very units only in so far as they form part of the entire structure of science, taken both in its theoretical and practical dimensions.

Such interaction between the whole and the constituent parts is often referred to by the term 'holism'. Given that there is no clear consensus as to the meaning of this term, I consecrate the first half of this book to a detailed investigation of holism when used specifically to refer to the complexity of science. The basic insight here is that, in science, we have an inevitable interaction not only between one thought and another, but also between one inquirer and another. In the second half of the book, I present one possible way of taking holism seriously in our understanding of science. This part of the book makes no pretension to be, even approximately, a complete treatment of the way holism could be integrated into a proper account of science. In spite of this lack of completeness, however, I hope that the work presented here will be useful not only for those wishing to clarify the meaning of holism, but also for those who think that a holistic approach to science undermines scientific realism. For the benefit of the latter, I make a special effort to situate the entire discussion within the framework of arguments for and against realism. An essential part of the purpose of this book is to show that scientific realism is still a defensible philosophical position even when a deliberate and serious attempt is made to understand science in all its complexity.

The project began as a Ph.D. thesis at the Department of History and Philosophy of Science, Cambridge University. Thanks are due to friends and colleagues who have generously read parts of the book and have given their advice. For such assistance, I am particularly indebted to Professor Nicholas Jardine, Professor Peter Lipton, and Dr. William Newton-Smith. I learned much from formal and informal discussions at the Department of History and Philosophy of Science, and also from the Tuesday meetings of the Moral Science Club, attended by many members of the Faculty of Philosophy. Thanks are due also to St. Edmund's College for accommodation and a friendly atmosphere during the early stages of my work. For the funding which made this research possible, I am grateful to the Pontifical Gregorian University of Rome, and to the Society of Jesus.

Chapter One

The Task Ahead

Holism is a broad philosophical position that has had applications in a vast number of research fields including such diverse areas as semantics, quantum mechanics and medicine. It can be described in general terms by the well-known assertion that certain wholes are greater than the sum of their parts. Since this description leaves a lot to be desired as regards precision, various philosophical studies in the course of history have been dedicated to clarifying the essential claims of holism in one way or another.[1] It seems undeniable that some composites are not just a collection of independent members, but seem to have an internal organisation that gives the composite something more than can be given by the sum of its parts. A living cell is quite different from a heap of sand. An essential claim of holism is that, on the basis of this observation, one should not engage in an investigation in a way that concentrates exclusively on small units with no reference whatsoever to the wholes they may be part of. Taking holism seriously means accepting that any one entity, even if it is a thought, a hypothesis, a theory, or an inquirer, should not be considered in isolation without any reference to the whole it may form part of. One needs to take into consideration its relation to other parts that make up the whole.

This attitude, resulting from the claim that some holistic systems do exist, is opposed to the attitude usually associated with the empirical method, in as much as this method, favouring analysis, is limited to consider wholes as simple aggregates of their parts. When one accepts that the object of an investigation is a part of a holistic composite, one needs to allow for the fact that the properties of such a part are determined to some extent by the relations it bears to other parts and to the whole composite itself. In this case, a systematic neglect of such relations will lead to an investigation that will be at best incomplete, or at worst incorrect.

To avoid such deformities in our understanding, therefore, one must ensure that the central working concepts employed at the very start of any investigation do not bring in with them a hidden assumption that the system one is dealing with is not

---

[1] For a survey of the historical roots and broad application of this philosophical position, three useful works are: D. Lerner, (ed.), *Parts and Wholes*; D.C. Phillips, *Holistic Thought in Social Science*; C. Lawrence and G. Weisz (eds.), *Greater than the Parts: Holism in Biomedicine 1920-1950*.

holistic. Such hidden assumptions, like the soldiers within a Trojan horse, will not be evident at the very start. The central concept will appear useful and impressive in its ability to simplify our investigation. The hidden, anti-holistic assumptions will, however, emerge later and end up by eventually opening the gates to all kinds of misunderstandings.

Each of these points certainly deserves more careful discussion, but this will be postponed to later sections of the book. It will here suffice to mention that, when trying to arrive at a correct understanding of a given thing, some key concepts employed may be more convenient than others in allowing for the possibility of holism within the system being dealt with. Such concepts that allow for holism will be called holistic concepts. In terms of this vocabulary, the main thesis to be defended in the following chapters can be summarised as follows: first, that the inclusion of holistic concepts in our understanding of science is essential, and, second, that an understanding of science resulting from the use of such holistic concepts adds considerable plausibility to scientific realism. Some preliminary remarks on realism will serve as a precious clarification of the general direction our investigation will be taking.

Realism is often presented in a variety of formulations, and one may legitimately wonder whether the different formulations are descriptions of a single philosophical position or whether there are various realisms.[2] In what follows, I will limit myself to realism as it concerns our understanding of scientific theories and practices, and I will furthermore assume that the philosophical position we refer to when using the term realism in this limited realm is in fact a single position. It can be briefly characterised by two claims: first that the external world exists, and second that we are capable of coming to know something about it through scientific inquiry. That these two claims are significantly different can be seen from the way realists struggle to respond to the sceptic who often accepts one but rejects the other. Such sceptics argue that, even though the external world is assumed to exist, we still cannot claim to know anything about it. Realists take the sceptical challenge seriously but are obviously not sceptics themselves. They want their task to be difficult to accomplish but not outright impossible. As Crispin Wright has elegantly put it, 'the unique attraction of realism is the nice balance of feasibility and dignity that it offers to our quest for knowledge ... We want the mountain to be

---

2 See R. Boyd, 'What Realism implies and what it does not'; 'The current Status of Scientific Realism'; J. Haldane and C. Wright (eds.), *Reality, Representation, and Projection*; G. Hellmann, 'Realist Principles'; P. Horwich, 'Three Forms of Realism'; H. Putnam, *Meaning and the Moral Sciences*; *Reason, Truth and History*; *Realism and Reason*.

climbable, but we also want it to be a real mountain, not some sort of reification of aspects of ourselves'.[3]

Realists are therefore committed to hold on to both claims at the same time. They consider scientific inquiry successful when it enables us to know something about the external world. This is taken to mean that scientific inquiry is successful when the picture of the world resulting from the inquiry is a good copy of the external world. It is successful when what the theory entails is true because it corresponds to the external world. For some realists, obtaining an exact copy of part of the external world appears too ambitious a task. Hence, they are satisfied with accepting one theory as better than another on the grounds that the first has more verisimilitude than the second. This means that it has more true logical consequences and less false ones than the second.[4] But even realists who opt for this less ambitious move cannot apparently escape having to do with correct or incorrect copies of the external world: the logical consequences are true or false according to whether they are good copies or not. The correspondence theory of truth has therefore been so intimately related to the realist position that many consider it constitutive of scientific realism.

But there are good reasons to make realists think twice before endorsing the correspondence theory of truth. It has its own inherent problems. The claim that truth is a relation of isomorphism between pieces of language and pieces of the world has a long history.[5] To understand what is at issue, we must be careful not to state mere platitudes. Protagonists of the correspondence theory want to claim more than merely that a true judgement gets things right, or that, to take Aristotle's classic expression, to speak truly is to say of what is that it is, and of what is not that it is not. If we are realists, we hold that the division between judgements which are true and those which are false is determined by something independent of language itself. Notice that there are two issues here. The first claim, namely that there is in fact a division between judgements which are true and judgements which are false, is a necessity of any comprehensible world. Accepting this does not make us protagonists of the correspondence theory of truth. It is the second issue which does this, namely the claim that we can explain, justify or legitimate such a division by alluding to a relation that the judgements have to something else. Scientific

---

[3] C. Wright, 'Realism, Antirealism, Irrealism, Quasi-Realism', p. 25.

[4] K. Popper, *Conjectures and Refutations*, p. 233; W. H. Newton-Smith, *The Rationality of Science*, chapter 8.

[5] D.J. O'Connor, *The Correspondence Theory of Truth*; N. Jardine, *The Fortunes of Inquiry*, pp. 10-13.

realists have often considered a *particular* relation as relevant, namely the one the judgements have to extra-linguistic objects and states of affairs.

But there are well-known problems. The correspondence theory claims that truth is a relation between what is said or thought and a fact or state of affairs in the world. Problems can arise when we ask questions about the nature of the first term of the relation and they arise as well when asking about the nature of the second. The first term assumes that one can easily identify what is said or thought. This is not so clear. What is said cannot be the sentence, because one thought can be described by different sentences, one in English, say, and its translation in French. What is common to all these equivalent sentences has been called a proposition. But the nature of such propositions is difficult to conceive. They must be timeless and wordless entities, if 'words' is understood in the usual sense.[6] So the least one can say at this stage is that the notion of truth-bearer is not straightforward. The second problem has to do with the second term of the relation of correspondence. There does not seem to be any means of grasping any part of mind-independent reality other than through linguistic representation. We cannot therefore conceive of a comparison between our representations and the world, a comparison which would enable us to assess the adequacy of our representations, their correctness or incorrectness. All our presumed commerce with that world is mediated by representations. We can compare our representations and judgements only with one another. The first problem plagues all theories of truth, while the second is particular to the correspondence theory. I will therefore concentrate mainly on this latter one. I will proceed by mentioning briefly two attempts that have been suggested to solve it.

The first consists in the claim that when employing the correspondence theory we are not doing commerce with the world at all but with linguistic conventions.[7] The belief that galaxies are in mutual recession owes its truth to the fact that galaxies are in mutual recession in the actual universe we live in. There are two kinds of linguistic conventions at work here. On the one hand we have descriptive conventions correlating the words with the *type* of situation, thing, event, and so on, found in the world. This gives us the facts. Hence an example of a type of situation is one which we conventionally take as a universe in which galaxies are in mutual

---

[6] Some philosophers take propositions to be sentences of a mental language common to all language users. I am not considering bits of this mentalese to be words, because my assumption is that for something to be a word there must be the possibility of it being uttered or written down.

[7] This position is based on the debate between J. L. Austin and P. F. Strawson. Cf. J. L. Austin, 'Truth'; P. F. Strawson, 'Truth'.

recession — as opposed to a universe in which galaxies behave otherwise. On the other hand, we have demonstrative conventions correlating the words with the historic, *token* situations, things, events and so on. In this case, we have a convention concerning these galaxies here, this behaviour here. Hence, given these two linguistic conventions, we can claim that what the correspondence theorist is doing is matching what language describes as types of things, situations, events, etc. with what language indicates as particular tokens of things, situations, events, etc.

But such a solution begs the question. It attempts to give an account of correspondence but in fact needs the notion of correspondence in its argumentation. To settle on a descriptive convention, that is to decide that such and such a sentence describes a certain type of situation, we seem to need some principle that tells us which of our descriptions does in fact give the *correct* description of these situations. In the example above, the descriptive convention is meant to pick out, uniquely and without ambiguity, a type of situation, namely a universe in which galaxies are in mutual recession. Not any descriptive convention would do. We have to use these specific words, and use them in this specific way. One can envisage perhaps a situation where a new terminology is introduced in a particular descriptive convention. But if we assume, like Austin, that descriptive conventions are completely arbitrary, then we are not doing justice to the fact that most of the words in our description already have a meaning before we use them. To pick on the *right* meaning, and hence to give a *correct* description, we inevitably have recourse to a kind of correspondence between what we suggest and what is the case. It is in this sense that the solution begs the question.

The second way to solve the problem involves an attempt to make the second term of the correspondence relation accessible. In other words, it is an attempt to make the comparison with the real non-linguistic world unproblematic. The basic idea here is to hold that what acts as arbitrator in our truth-value deliberations is not the non-linguistic world at all but the ultimate scientific theory, understood here as the upshot of indefinitely prolonged scientific inquiry, the theory that would be converged on were human inquiry engaged in indefinitely. Accordingly, it is reasonable to say that now a given statement is true in the case when we have good reasons to believe that it will withstand all future criticism and hence will be included in the ultimate theory. It will be one of the features of the conceptual scheme on which human inquiry in ideal conditions will converge upon in the limit. This view of things can be considered successful in solving the correspondence problem mentioned above because we now apparently have a much clearer view of what we are comparing to what, and in what sense. The first problem is however immediately evident. Is the ultimate theory conceivable? Some attempts have been made to explicate the ultimate theory in terms analogous to numerical convergence of series, but it is generally agreed that such elaborations lose much of their explaining power when one realises that the kind of closeness of one theory and a successor theory in one area of science has little in common with the kind of closeness of one theory and its successor in another area of science.[8] Moreover, the idea of a theory which satisfies, and is known to satisfy, all possible standards of acceptability presupposes the idea of the totality of such demands. And this latter idea cannot be given any content. In fact, the fixing of these standards is probably undertaken to a certain extent in the very process of theorising. This solution to the correspondence problem thus starts looking precarious.

Even if one concedes the conceivability of the ultimate theory, the correspondence problem is not solved by the introduction of the notion of the limit of inquiry. The main question that brings out the dilemma is the following. In this new understanding of correspondence-truth, what kind of relation is being presupposed to hold between the ultimate theory and the non-linguistic real world? There are two alternatives. One is to hold that there is nothing more to this non-linguistic real world than that which is portrayed by the ultimate theory. Hence on this alternative, there is total identification of real world and ultimate theory. The other alternative is to hold that the non-linguistic real world is that which corresponds to the ultimate theory. On this alternative, the decisive criterion of truth is the real world.[9] But both these alternatives are considerably problematic. The first one introduces idealism of an extreme kind difficult to accept even if we suppose that theories do converge on an ultimate one. The second begs the question because it employs the very same concept of correspondence it was meant to explicate in the first place.

The least we can say is that the problems of the correspondence theory are resistant. It is understandable therefore why scientific realists should seek to give an account of their position without having to endorse the correspondence theory. I will be arguing that the fact that the correspondence theory of truth is very vulnerable does not mean that scientific realism is also insecure. An understanding of scientific activity in line with realism can indeed be formulated, on condition that our reflections have a different starting point from that of the received view discussed so far.

---

[8] N. Jardine, *Fortunes of Inquiry*, chapter 2. I am also indebted to C. J. Misak, *Truth and the End of Inquiry*.

[9] The first alternative is taken by J. Rosenberg, *Linguistic Representation*; the second by W. Sellars, *Science, Perception and Reality*.

To get an idea of what this starting point is, consider one important central concept used extensively in discussing science: the notion of 'theory'. Most studies include discussions concerning the relation of one theory to another, the formulation of new theories, the confirmation or falsification of theories, the progression towards an ultimate theory, and so on. But to describe to a reasonable degree of precision what a scientific theory is seems to be a difficult task. An attempt has been made by W. H. Newton-Smith in the following words: 'By a *theory* I shall mean the deductive closure of a set of theoretical postulates together with an appropriate set of auxiliary hypotheses; that is everything that can be deduced from this set.'[10] It is true that, in textbooks, theories are often presented as well-defined units in the way suggested by Newton-Smith. But such a definition, although good enough for the purposes Newton-Smith has in mind, does not emphasise the inevitable links between theories. Even the simplest cases, like the kinetic theory of gases, are not as self-contained as the definition seems to suggest. The kinetic theory is a term applied to the study of gases by assuming that gases are composed of molecules subject to rapid random motions colliding with one another and with the walls of the container. However, for the analysis of these motions, which is part of the theory itself, one must employ concepts and methods pertaining to the domain of mechanics.[11] If one concedes that individuating is possible, then one must apparently also admit that any given theory is linked to others in a chain-like structure in such a way that if one theory sinks, it brings down others with it. If one theory is falsified, all the other theories using any one of its conclusions as a basis for an auxiliary hypothesis, or as a basis for its central concepts, are put into question. It is very probable that distortions in our understanding of science will occur if such dependence of one theory on another is systematically avoided. One conclusion to be drawn therefore is that our understanding of science is better approached from the starting point that allows for the possibility of holism, rather than from an analysis of isolated units assumed to be well-defined on their own without reference to others.

A tendency to understand science through holistic concepts can be discerned in some influential studies. A kind of sequence can be observed involving central concepts that become more and more holistic. A very influential account has the well known starting point that a universal theory can be disproved by only one

---

[10] W. H. Newton-Smith, *The Rationality of Science*, p. 199.

[11] In fact, one of the more disputed aspects of the kinetic theory is the notion employed in mechanics of equipartition of energy, according to which each degree of freedom of the gas possesses the same amount of energy. Cf. R. N. Varney, 'Kinetic Theory'.

counter-instance of it.[12] Protagonists of this account advocate a method of conjecture and refutation according to which one should consciously seek theories which are falsifiable, and then test them so that those which survive are tentatively accepted as being closer to the truth. Hence, as regards the epistemology involved in science, this account is focused mainly on *theories*. A second account starts by considering scientists working within a paradigm.[13] Here, if a puzzle facing researchers within a paradigm proves itself too stubborn, it becomes an anomaly. The researchers will then be in a crisis state, where normal science is abandoned for extraordinary science. The anomaly often leads to the proposal of a candidate for a new paradigm, and if this new paradigm wins the debate, a scientific revolution occurs. Hence, one can see that, on this account, one major, central concept is that of *paradigm*. A third influential account introduces the idea of research programmes which are described as having two major parts: the hard core and the protective belt consisting of auxiliary hypotheses.[14] When the predicted novel facts do not receive confirmation, a research programme starts to degenerate. Prolonged lack of confirmation shows that the protective belt is losing its function. However it is not sufficient to eliminate a programme solely on the basis that it presently appears not to be making empirical progress. The programme could be in a degenerating phase from which it will recover. Hence the competition between rival research programmes can continue for centuries, because there is no such thing as a crucial experiment for a final decision. In this account, the central concept at work is that of *research programme*. A more recent account presents science through a central concept referring to the set of general assumptions about the allowed ontology and methodology.[15] It is a set of assumptions concerning the entities and processes in the domain of study, and concerning the appropriate methods to be used for investigating the problems and constructing the theories in that domain. This set of assumptions is called a *research tradition*. From these four accounts, one can discern some traces of a trend to give an account of science by using central concepts that are designed to do more justice to the complexity and mutual dependence of various aspects of science: a tendency to employ more holistic central concepts. As regards work in the history of science and in the sociology of science and of scientific knowledge, concepts meant to emphasise holism are used

---

[12] K. Popper, *Conjectures and Refutations: the Growth of Scientific Knowledge*.

[13] T. Kuhn, *The Structure of Scientific Revolutions*.

[14] I. Lakatos, *The Methodology of Scientific Research Programmes*.

[15] L. Laudan, *Progress and its Problems*.

very frequently but are not accompanied by any kind of philosophical analysis about their correct use or possible misuse.

My aim is to engage in such a philosophical analysis. I will be concentrating on the role of central holistic concepts in the understanding of science taken as a whole. The term holism, therefore, will not be referring primarily to the understanding of material composites and how these behave or function. It will not be referring primarily to the understanding of particular systems studied by scientists, like, say, the understanding of biological systems, or the understanding of molecules with emergent properties not found in their constituent atoms. Even though the relevance of holism in such areas of investigation is important, it will not be the main point of interest in the following chapters. The emphasis will be rather on the understanding of science itself. In other words, I will not be situating myself within science and engaging in a reflection on how holism changes the work of scientists in one particular field of inquiry, like in the area of biology, or of quantum mechanics. I will be engaging in a meta-discourse, a philosophical reflection that applies holism to the understanding of science itself. The composite in consideration will therefore be the entire system we call science.

One important consequence of adopting this strategy is that the ensuing investigation will be simultaneously related to what are customarily considered different areas of science-studies. Hence, when holism is applied in this way to the entire system of science, considerable work has to be done in relation to theories of meaning, in relation to the question of change and permanence in the course of history, and also in relation to how one should understand the terms 'practice' and 'discourse'. One is touching here, therefore, not only major issues dealing with the account of science usually associated with analytic philosophers, but also major issues within the account of historians of science, and major issues within the account of sociologists of science. In bringing together these various accounts of science, a consideration of holism in the understanding of science will be a step towards appreciating the mutual enrichment that these various disciplines can give rise to. This is beneficial especially because these three major areas of science-studies have tended, during the last decades, to trace their own paths and diverge from each other.

As was mentioned earlier, the main thesis to be defended can be briefly described in two points. The first one is that the inclusion of holistic concepts in our understanding of science is essential. This will take up the first half of the book. The second half will be dedicated to the second point, namely that the account of science resulting from the use of such holistic concepts adds considerable plausibility to scientific realism.

Chapters two and three will be dedicated to the first point of the thesis. One will be about cognitive aspects involving the dependence of one thought on other thoughts. The other chapter will be about social aspects involving the dependence of one inquirer on other inquirers. In chapter four I will concentrate on two holistic concepts which are often used in philosophical, sociological and historical studies of science, namely the notion of discourse, which is more related to cognitive holism, and the notion of practice, which is more related to social holism. The use of these concepts guarantees that holism will not be neglected in our understanding of science. In chapter five I will apply the general conclusions of chapter four to the case of science. My special aim here will be to defend the claim that scientific practice can be viewed, in some respects, as a single whole, as opposed to the claim that science is made up of unconnected local practices. It will be shown that the inclusion of holism in our understanding of science obliges one to explore the relation between scientific practice and common practice. In chapter six I investigate the latter kind of practice. In broad terms, common practice will be understood as that practice in which common sense is embedded. In chapter seven, I will concentrate on what constrains practices. I will argue that the *kind* of constraints acting on practices can be used to give an account of objectivity as an attribute not of judgements but of practices. The final chapter will then contain an argument showing that scientific practice, in the terms developed in the preceding chapters, can be considered highly objective and self-adjusting. In this way, scientific realism will be shown to be plausible even though the validity of the correspondence theory of truth is not presupposed.

Chapter Two

Cognitive Holism

One way of endorsing holism within a philosophical investigation is to hold, at the very least, that concentrating exclusively on small units with no reference whatsoever to the wholes they may be part of, should be avoided. Taking holism seriously means accepting that any one entity, part of a whole, should be considered in relation to other parts that make up that whole. Endorsing holism in the understanding of science is one particular application of this general proposal, and the entities concerned in this case will be of various kinds, for example, thoughts, theories, purposes, experiments, inquirers, and even scientific institutions.

To realise why the question of holism, in some form or other, merits critical attention from those wishing to gain a deeper understanding of science, it is enough to recall that a discipline like science, although it covers only a limited area within the vast stretch of human intellectual activity, is a very complex enterprise, involving not only purely theoretical considerations but also interactions between individuals and between groups, interaction resulting sometimes in agreement, sometimes in disagreement. Although it may often be easier to postulate a clear distinction between theoretical conjectures and the practical involvement of scientists who make them, one has to admit that the actual world we live in does not come ready-made with any such clear distinctions. There is certainly some interdependence between one concept and another, between one inquirer and another, and between concepts and inquirers.

Moreover, even if we concede that a division can plausibly be said to exist between the practical side and the theoretical side of science, we will still find, when limiting our considerations to the theoretical side only, that the investigation of parts cannot be undertaken without some reference to the larger wholes they are in. For example, some may hold that the scientific concept of mass is operationally definable in terms of physical and mathematical operations that can be carried out in the laboratory or in the field. On this view, the concept of mass is therefore identical to the description of an operation involving a particular experimental set-up: in this case, perhaps a beam-balance. But such a view is questionable. If we say that the scientific concept of mass is the same thing as the description of an experimental operation, then we will not be allowing the possibility of discrepancy between what we want to measure and what we actually measure. We will not be allowing two simple questions to be asked. Firstly, is this experimental operation involving a beam-balance really measuring mass rather than something else? Secondly, is this particular beam-balance working properly rather than wrongly? These are legitimate questions. We should therefore look for an account of the meaning of scientific terms which allow them. A more plausible explanation of what is happening in the example is the following: we assume we measure mass by using the beam-balance because we presuppose not only that there is a special relation between the term 'mass' and the particular operation involving that set-up, but also that the beam-balance is governed by some relevant physical laws. In this case, we have the laws of rigid-bodies and of conservation of angular momentum. If we use a beam-balance made of rubber, we will not say that we are measuring mass. The concept of mass therefore is identifiable not by an operation involving the experimental set-up but by the laws that that set-up is representing, here the laws of mechanics. The concept is not isolatable. It must be grasped within a network of concepts structured together constituting a law.[1]

But saying that a holistic outlook is preferable is not saying exactly what is involved by adopting it. Many versions of holism have been suggested and not all, it seems, are supported by cogent arguments. The example given above suggests that one way of giving a plausible account of holism in science is to emphasise the similarity between a number of relations. There is the relation between concepts and laws, where laws are considered larger units consisting of concepts. There is the relation between laws and theories, where theories are larger units of laws. There is the relation between theories and paradigms, where paradigms are larger units of theories, and so on. C. U. Moulines has provided a general definition of holism based upon such relations. He first defines scientific constructs. He holds that scientific constructs are conceptual units appearing in scientific expositions such as the concept of mass, or the law of gravitation, the theory of classical mechanics, and so on. In the general sense, therefore, holism will be the position according to which 'the proper working of a scientific construct A in context C presupposes the (proper) working of scientific construct B in C, and B is an integrating whole with respect to A'.[2] This definition can be understood better if an example is considered. One can take A to be the concept of mass, B can be taken to be Newton's Second Law of Motion, and C can be taken to be Newton's Theory of Motion. Moulines' definition of holism is saying that the proper working of the concept of mass in the Newtonian Theory presupposes the proper working of

---

[1] This example is from C. U. Moulines, 'The Ways of Holism'.

[2] *Ibid.*, p. 314.

Newton's Law in the theory, and Newton's Law is an integrating whole with respect to the concept of mass. From this definition, holism implies that, to arrive at the identity-conditions of a given scientific construct, one has to consider its location within a larger unit. If we seek the identity-conditions of a concept, we will need to consider it as a constituent of an empirical law. If we seek the identity-conditions of a law, we will need to consider it as a constituent of an empirical theory, and so on.

This suggestion by Moulines certainly gives a valuable preliminary idea of the relevance of holism, but it needs more elaboration. One aspect that makes such elaboration necessary is that no reference is made to recent work in semantics and cognitive science where the notion of holism has been analysed and discussed in a number of ways. In fact it is not difficult to see one possible connection between the suggestion made by Moulines and theories of meaning: for example, knowing the identity-conditions of a concept and knowing its meaning are intimately related, to say the least.

So what I propose to do in this chapter is first to clarify further some prominent versions of holism and to bring out the areas where the account of holism usually associated with semantics is relevant to the holism associated with our understanding of science. The second section will include an evaluation of a major objection that applies to all types of holism. In the third section I will take up the challenge of this objection and show how a minimal form of holism may nevertheless be vindicated. In this chapter I will be concentrating on aspects that relate to cognition rather than on aspects that relate to social groupings of inquirers. This social aspect of holism will be discussed in the next chapter.

1. Ways of Understanding Holism

Confirmation Holism

Confirmation holism is one of the widely discussed kinds of holism in the domain of philosophy of science. My aim is to arrive at a reasonably clear idea of what confirmation holism is. A good starting point is Pierre Duhem's work on the structure of physical theories. The account of holism described here was extended in a significant way by Willard van Orman Quine, especially in his important paper 'Two Dogmas of Empiricism'. The resulting account of holism, which has come to be known as the Quine-Duhem thesis, has been applied to the philosophy of mind by Donald Davidson. Consideration of the works of these three authors will enable me to arrive at a reasonably clear account of confirmation holism.

Duhem's original argument is essentially an inductive argument based on a number of historical examples. These examples give reasons to believe the general claim that a scientific hypothesis cannot be isolated and tested on its own. Duhem's concluding statement, which is a good description of confirmation holism, goes as follows:

> In sum the physicist can never subject an isolated hypothesis to experimental test, but only a whole group of hypotheses; when the experiment is in disagreement with its predictions, what he learns is that at least one of the hypotheses constituting the group is unacceptable and ought to be modified; but the experiment does not designate which one should be changed.
>
> We have gone a long way from the conception of the experimental method arbitrarily held by persons unfamiliar with its actual functioning. People generally think that each one of the hypotheses employed in physics can be taken in isolation, checked by experiment, and then, when many varied tests have established its validity, given a definitive place in the system of physics. In reality, this is not the case. Physics is not a machine which lets itself be taken apart; we cannot try each piece in isolation and, in order to adjust it, wait until its solidity has been carefully checked. Physical science is a system that must be taken as a whole ...[3]

So when we want to revise our theories because of some surprising observation, we may be left with a choice between different revisions which would all restore consistency. It is not always clear whether Duhem is taking the whole of physics to be implied in the confirmation of a given hypothesis, or just a group of neighbouring theories. Whatever the answer to this question, it seems clear that he did not intend his confirmation holism to go beyond the discipline of physics. In particular, he explicitly states elsewhere in his study of physical theories that there is a clear distinction between scientific claims and ontological ones.

It will be useful to mention that Duhem's thesis about confirmation effectively restores the symmetry between verification and refutation of theories. By not taking Duhem's thesis into consideration, one may hold that, from a strictly logical point of view, refutation of a theory is conclusive. If a theory T entails observational consequences O, then the falsity of T is deductively inferable by modus tollens

---

3 P. Duhem, *The Aim and Structure of Physical Theory*, p. 187. See also A. Grünbaum, 'The Falsifiability of Theories: total or partial? A Contemporary Evaluation of the Duhem-Quine Thesis'.

from the truth of the conjunction $(T \rightarrow O)\&\sim O$. But, as far as confirmation is concerned, the truth of T does not follow deductively from the truth of the conjunction $(T \rightarrow O)\&O$. So the confirmation of a theory is not conclusive. In other words, on neglecting the Duhem thesis, we have a situation where there is asymmetry between verification and refutation. On accepting it, symmetry is restored. Duhem claims that we are mistaken when we think that the observational consequences of a given hypothesis H are deduced from the hypothesis alone. Observational consequences are deduced from the conjunction of H and the relevant auxiliary assumptions A. This results in a situation where the refutability of H is no more conclusive that its verifiability. The truth of H cannot be deductively inferred from the truth of the conjunction $[(H\&A) \rightarrow O]\&O$. And the falsity of H likewise cannot be deductively inferred from the truth of the conjunction $[(H\&A) \rightarrow O]\&\sim O$.

This restored symmetry between verification and refutation is one of the major building blocks Quine uses in order to expound his views on the growth of knowledge in the last sections of his paper 'Two Dogmas of Empiricism'. On the basis of Duhem's arguments, Quine is convinced that 'our statements about the external world face the tribunal of sense experience not individually but only as a corporate body'.[4] In the final section, Quine brings out some of the consequences of his holistic empiricist view by suggesting that the totality of our knowledge is a man-made fabric which impinges on experience only at the edges. The total structure of our knowledge is underdetermined by our experience in such a way that 'any statement can be held true come what may if we make drastic enough adjustments elsewhere in the system'.[5] Such adjustments are apparently constrained not only by experience but also by what he calls 'our natural tendency to disturb the total system as little as possible'.[6]

It will be useful at this point to mention a way of applying Quine's insights to another very important branch of philosophy, namely the philosophy of mind. If instead of theories in general we have one's own personal theories about the beliefs

---

[4] W. V. O. Quine, 'Two Dogmas of Empiricism', p. 41; see also *Pursuit of Truth*, chapter 1; J. D. Greenwood, 'Two Dogmas of Neo-Empiricism: the "Theory-Informity" of Observation and the Quine-Duhem Thesis'; R. Klee, 'In Defence of the Quine-Duhem Thesis: a Reply to Greenwood'; C. Hookway, *Quine*.

[5] W. V. O. Quine, 'Two Dogmas of Empiricism', p. 43.

[6] *Ibid.*, p. 44.

of another person, and instead of data in general we have observable behaviour patterns of the other person, then we arrive at a version of Davidsonian holism.[7]

First, a word about the context: at one point in his writings, Davidson argues that there cannot be strict psychophysical laws.[8] By psychophysical law here is meant a law which connects mental states and events, like beliefs, desires, perceptions and so on, to physical states and events. The argument starts with some important observations about the intelligibility of measurement in the physical sciences. One can make sense of measuring length only if one situates the concept of length within the appropriate conceptual environment. This is the same point mentioned at the beginning of the chapter. There, it was briefly shown that we assume we measure mass by using the beam-balance, rather than measuring something else, because we presuppose that the beam-balance is governed by certain physical laws. Davidson is making a similar, more general point. He generalises by holding that the entire set of axioms, laws, or postulates for the measurement of mass, and of length, is partly constitutive of the idea of a system of macroscopic, rigid, physical objects, having identity over time.

This line of argument can be applied to the attribution of beliefs. We cannot assign a length to an object unless a certain coherent theory holds of objects of that kind. Similarly, we cannot attribute a belief to a person except within a framework of a viable theory of the person's overall beliefs and desires, intentions and so on:

> There is no assigning beliefs to a person one by one on the basis of his verbal behaviour, his choices, or other local signs no matter how plain and evident, for we make sense of particular beliefs only as they cohere with other beliefs, with preferences, with intentions, hopes, fears, expectations and the rest.[9]

When faced with a person's behaviour pattern, I do not identify propositional attitudes, and attribute them to the person, one by one without reference to each other. I make sense of particular propositional attitudes only as they cohere with

---

[7] What I am calling Davidsonian holism here is limited to his position concerning psychophysical laws. It does not refer to his semantic developments of this idea. According to the semantic version, one can only interpret a single sentence of the object language by means of its T-sentence if one is in a position to interpret many sentences in the same language. Being able to interpret a sentence is not simply knowing the appropriate T-sentence. See D. Davidson, 'Semantics for Natural Languages'.

[8] D. Davidson, 'Mental Events'. See also S. Evnine, *Donald Davidson*.

[9] D. Davidson, 'Mental Events', p. 221.

other propositional attitudes. The content of a propositional attitude derives from its place in the pattern.

From this point, Davidson draws the conclusion that mental and physical schemes are basically different. In brief, he wants to say that, on the one hand, physical change can be explained by laws that connect it with other changes and conditions physically described. On the other hand, the attribution of mental phenomena must be responsible to the background of reasons, beliefs and intentions of the individual. The two schemes therefore have different commitments. The conclusion for Davidson is that there can be no psychophysical laws.[10]

The plausibility of this argument is not my main concern. The point I want to underline is that Davidsonian holism is a special case of Quinean holism. Two points seem to bring this out most clearly.

It was shown above how Quine, following Duhem, argues that there is no correlation between single sentences and experiences which confirm of disconfirm them. Instead, one must hold that statements about the external world face the tribunal of sense experience not individually but only as a corporate body. Davidson is applying this point to the debate about mental states: instead of Quine's statements, he puts mental states; instead of Quine's sense experience, he puts behaviour. Hence, Davidson's holism can be expressed by the following claim: there is no element of behaviour of a person that obliges an observer to attribute to that person one particular mental state rather than another. Just as statements about the external world face the tribunal of sense experience as a corporate body, so also attributions of propositional attitudes face the tribunal of behaviour patterns as a corporate body.

Quine also made the point that every statement can be considered true come what may, if we make drastic enough adjustments elsewhere in the system. This point also has a counterpart in Davidson's version of holism. An example is the situation where what I observe of my friend's behaviour makes me spontaneously attribute to him a certain belief. Suppose that I then observe a new element of behaviour that seems to warrant my withdrawal of the attribution of that belief. Hence, I may have thought that he believed that giving alms is good. Then, to my

---

[10] This point is only half of the thesis Davidson defends as regards the mental. His general thesis is called the Anomalism of the mental, by which he means that mental states and events are not subject to scientific laws. Mental states and events are subject neither to *psychophysical* laws relating mental states and events to physical states and events, nor to *psychological* laws relating mental states and events to other mental states and events. In my argument, I draw attention only to the point about psychophysical laws.

surprise, I see him walk indifferently past a homeless person asking for change. Should I stop attributing that belief to him? The question has no clear answer. He can still be attributed the belief that giving alms is good, because he may also be attributed other beliefs. He may also have beliefs about homeless persons in general, for example the belief that they will not use the money properly if given to them in the streets, or he may have beliefs about that particular homeless person, for example the belief that that particular person is a swindler. Hence any belief attribution can be considered appropriate if we are ready to make drastic enough adjustments to our attributions of other propositional attitudes.

So, to recapitulate briefly, confirmation holism as discussed by Duhem, Quine and Davidson can be described as the philosophical position according to which no observation can confirm a single statement or hypothesis taken in isolation. A whole group of hypotheses is involved each time, and a recalcitrant observation does not tell us which hypothesis should be changed.

Semantic Holism

Semantic holism can be conveniently introduced by recalling that, in the first sections of Quine's paper 'Two Dogmas of Empiricism', we find a number of arguments concerning the distinction between analytic and synthetic statements. In general terms, I will be following Quine by taking statements to be analytic when their truth is grounded on meanings independently of matters of fact. Statements are synthetic when their truth is grounded on fact. So, although Quine does agree with and substantiate the Duhem thesis of holism as mentioned above, he does not do this in a way unrelated to semantics, as the brief discussion above may suggest, but within the context of a debate concerning issues previously discussed by Rudolf Carnap. It is by discussing Duhem's ideas within this kind of context that Quine makes a valuable contribution to the understanding of holism.

Carnap opposed the claim of Frege and Russell that the logic they had developed was in fact the common framework of all languages. To remedy this and be more in line with Duhem's holism, Carnap envisaged a plurality of linguistic frameworks, each containing a system of logical principles. The terms employed in each framework obtain their meaning by analytic principles linking the claims employing them to observational claims. For example, one of the analytic principles determining the meaning of 'acid' might be: 'a substance is an acid if it turns litmus paper red'. Hence, for Carnap, if inquirers share a framework, they can arrive at objective answers to their questions. But disputes about which framework to employ cannot be settled by saying that one of the given frameworks is true or false. The choice can only be judged as being more or less expedient, fruitful, and conducive to the aims of the language-users.

Carnap uses Duhem's ideas on holism to make the claim that, when faced with observational surprise, we have a choice between different revisions of our system of hypotheses, all of which restore consistency. When this is viewed in terms of linguistic frameworks, however, he made it clear that, while an observational surprise may conflict with a system of statements, it cannot conflict with an analytic proposition. Within one framework, an analytic proposition is totally invulnerable. What may happen, however, is that, under the inducement of a great number of observation sentences, we change the linguistic framework for pragmatic reasons. We can change it to such an extent that a previously analytic sentence will no longer be analytic.[11] This picture therefore results in the situation where analytic statements can be used to identify a linguistic framework. If two inquirers have different linguistic frameworks, we should be able to identify these frameworks by noticing which of the assertions they make express analytic truths and which are synthetic, that is confirmed relative to the framework. Carnap therefore gives a specific role to the analytic-synthetic distinction.

This is the position Quine is reacting against in 'Two Dogmas of Empiricism'. Although Quine continues in the same direction as Carnap in the application of Duhem's holism, he does not agree with Carnap that such an application leaves room for a special role to be played by the analytic-synthetic distinction.

Quine evaluates a number of ways we can understand analyticity. In the first section he distinguishes between two types of analytic statements. The first type consists of logical truths, like 'no unmarried man is married'. These statements are true whatever meaning we give to 'man' and to 'unmarried'. The second type of analytic statements include those which depend on the notion of synonymy, like 'no bachelor is married'. Here the first logical truth is rewritten in a new version on the assumption that 'bachelor' and 'unmarried man' are synonymous. In the next two sections, Quine shows that analytic statements of the second type are problematic because the notion of synonymy is problematic. It cannot be explained by saying that A is synonymous to B if and only if A is defined as B. The very notion of definition depends on prior relations of synonymy. Nor can it be explained by holding that the synonymy of two linguistic forms consists simply in their interchangeability in all contexts without change of truth value. Quine shows that a language which allows such interchangeability *salva veritate* presupposes analyticity and thus begs the original question. Synonymy is therefore not useful to

---

[11] R. Carnap, *The Logical Syntax of Language*, pp. 318-319.

define analyticity. Suppose we try to understand analyticity by construing some kind of artificial language. This can provide some hope of defining the notion in terms of rules within that language. But Quine shows that, even here, such a procedure presupposes the very concept we are trying to understand. So all attempts to clarify the notion of analyticity are as suspicious as the notion itself.

The only way to save the analytic-synthetic distinction would seem to be to give it the role Carnap had given it, namely the role of identifying the linguistic framework within which a given inquirer is situated. But the explanatory work done by drawing a sharp distinction between analytic and synthetic statements is, according to Quine, null. To show this, he describes how we often start from the observation that the truth of a statement depends on both language and extralinguistic fact and then slide into the error of thinking that such a sharp distinction can be drawn.

> The statement 'Brutus killed Caesar' would be false if the world had been different in certain ways, but it would also be false if the word 'killed' happened rather to have the sense of 'begat'. Thus one is tempted to suppose in general that the truth of a statement is somehow analysable into a linguistic component and a factual component. Given this supposition, it next seems reasonable that in some statements the factual component should be null; and these are the analytic statements. But, for all its a priori reasonableness, a boundary between analytic and synthetic statements simply has not been drawn.[12]

Suppose we consider the basic belief that fire burns. One inquirer may say that this forms part of our concept of fire and of burning even though it is not one of the analytic principles of our framework. This inquirer will insist that we rely on it for predictions but were it to become questionable, even though we cannot now imagine how that could come about, we would be ready to adjust it. As opposed to this inquirer, another inquirer may be a follower of Carnap and perhaps claim that the sentence is analytic. Hence now, the truth of the claim 'fire burns' depends entirely on the meaning of the words. Within the inquirer's framework, such an analytic proposition stands unchanged in the face of observational surprises. However, this second inquirer, to be in line with Carnap's view, would also hold that the inducement of these observational surprises may be so great that pragmatic reasons will dictate a change of linguistic framework. The new framework may be such that, within it, 'fire burns' is no longer analytic.

---

[12] W. V. O. Quine, 'Two Dogmas of Empiricism', pp. 36-37.

Now compare the two inquirers. Is anything being added to our understanding of the relation between evidence and belief by adopting the second inquirer's complexities rather than the first inquirer's simpler picture? Quine insists that there is nothing being added. The distinction between analytic and synthetic statements has no significant explanatory role to play. In fact, one can even say that the explanation involving the distinction is unnecessarily awkward and complicated.

Some critics have insisted that Quine overlooked certain cases where analyticity is obvious: for example when we have statements of the form 'A is B' and we have no other way of understanding A except this formula.[13] But, if the above exposition of his ideas is correct, what he is claiming is that the *distinction* between analytic and synthetic statements is not useful in the kind of explanation of meaning, rationality and knowledge that Carnap desired. It is better to say that there is no clear boundary — that analyticity, or indeed syntheticity, is a matter of degree, all the way down. What we need for mutual understanding is that we are confident that others will, very often, apply terms to new cases as we do. We all know that lemons are normally yellow when ripe. We may loosely call this a case of analyticity. However, the statement 'lemons are normally yellow when ripe' can be called analytic simply because everyone expects everyone to accept it, and because it is taught as people learn their language. We tend to think of anyone who denies it as someone who does not understand the word 'lemon'. Such statements, often called analytic, can equally be called well-established synthetic ones.

If the analytic-synthetic distinction is abandoned, adjustments to our understanding of meaning must take place. In 'Two Dogmas of Empiricism', Quine makes the suggestion that, if we take the Duhem thesis seriously, then to speak of a linguistic component and a factual component in the truth of any *individual* statement like 'Brutus killed Caesar' is incorrect. The double dependence on language and experience has to be attributed to the whole of science. If we take the *statement* as the unit of meaning, we would be drawing our grid too finely. Hence, his suggestion is that 'the unit of empirical significance is the whole of science'.[14]

---

13 See H. P. Grice and P. F. Strawson, 'In Defence of a Dogma'.

14 W. V. O. Quine, 'Two Dogmas of Empiricism', p. 42. In later writings, Quine tones down this statement: 'Should it be the whole of science or the whole of *a* science, a branch of science? This should be seen as a matter of degree, and of diminishing returns. All sciences interlock to some extent; they share a common logic and generally some common part of mathematics, even when nothing else. It is an uninteresting legalism, however, to think of our scientific system of the world as involved *en bloc* in every prediction. More modest chunks suffice, and so may be ascribed their independent empirical meaning, nearly enough, since some vagueness in meaning must be allowed for in any event.' W. V. O. Quine, *Theories and Things*, p. 71.

These considerations suggest the following three remarks. First, one important consequence of endorsing the Duhem thesis is that the analytic-synthetic distinction has no significant role to play in our understanding of how knowledge grows. Secondly, when our centre of attention is not on the question whether parts of a theory are confirmed or not but on the question whether parts of a theory have meaning or not, then what we are dealing with is semantic holism rather than confirmation holism. Third, semantic holism holds that what serves as the units of meaning are extended parts of the theory, or extended parts of the discipline, or ultimately the whole our knowledge, or culture.

This semantic holism and the confirmation holism discussed earlier, although related, cannot easily be brought together to give a unified version of holism. I will mention two major problems. First of all, one may doubt whether confirmation holism and semantic holism actually concern the same entities.[15] The former holism concerns entities which can be confirmed or falsified. Hence, it does not concern sentences, which can be in one language or in another. It concerns what these sentences are about. Hence one does not confirm the sentence 'galaxies are in mutual recession' but *that* galaxies are in mutual recession. Confirmation holism thus concerns propositions, which are trans-linguistic entities. Semantic holism, on the other hand, concerns entities which can bear meaning. These entities are sentences or formulae or other entities that are essentially linguistic. Hence one discusses the meaning of the sentence 'galaxies are in mutual recession' and compares it with the meaning of the sentence 'les galaxies se reculent mutuellement'. Ambiguity therefore creeps in when one tries to combine semantic holism and confirmation holism in some kind of quick overall view, because the two holisms concern different things.

The second source of worry involves not the relation between confirmation holism and semantic holism but the very foundations of these kinds of holism, especially the first one. We have started with Duhem's idea of confirmation holism, which, as we have seen, can be formulated as the claim that it is always possible to save a given hypothesis by revising some of the background beliefs. Moreover, accepting semantic holism obliges us to claim, if we follow Quine, that the unit of meaning is the entire system of beliefs. But it is apparently arguable that the formulation of confirmation holism is dependent on the possibility of individuating definite pieces of our entire system and calling them beliefs, or hypotheses. How can we pick out such definite pieces of our entire system? Quine's suggestion was that meaning is an attribute of the entire system. We can perhaps try to weaken this

---

15 This criticism is from J. Fodor and E. Lepore, *Holism, a Shopper's Guide*, pp. 37-54.

suggestion by saying that meaning is not exclusively an attribute of the entire system, but can be an attribute of smaller units on condition that the entire system is involved: in other words that meaning is not global but globally determined. But, if we take Quine's suggestion at face value, we will be obliged to accept the stronger interpretation and hold that nothing can properly be said to have meaning except the entire system of beliefs. And this would entail that the usual way of individuating a belief by distinguishing its meaning from that of others is not available anymore.

Attribute Holism

These ambiguities encourage one to attempt to formulate the thesis of holism from a completely different angle. One such recent attempt involves taking holism to be a second-order attribute: a property of properties.[16] One of the merits of this approach is that it makes us focus our attention less on the abstract substantive 'holism' and more on the adjective 'holistic', and thus makes the entire issue more precise.

A *holistic* property is a property such that, if one thing has it, indefinitely many other things have it as well. Otherwise, the property will be non-holistic: either *atomic*, in the sense that if something has it, only that thing has, or *anatomic*, in the sense that if something has it, at least one other thing has. Hence, being a natural number is a holistic property. This is so because numbers are usually defined by reference to a successor relation. A number is neither its own successor nor its successor's successor. Hence the determination of each natural number is dependent on the determination of other natural numbers in such a way that, if there is any number at all, then there must be an infinity of them. For an example of an atomic property, consider 'being the oldest in a particular group'. This property has only one instantiation, assuming of course that there are no twins in the group. An example of an anatomic property is 'being a sibling'. There cannot be just one instantiation of this property, because if I am a sibling then there must be someone else whose sibling I am. It is useful to notice here that these three types of properties are not exhaustive: for example, the property 'is a New Testament scholar at Cambridge' is neither holistic, nor atomic, nor anatomic. The fact that someone has this property says nothing about whether others should have it as well.

One may use this definition of holism in terms of a distinction between properties to analyse and discuss the previous accounts of holism and also possible

---

16 J. Fodor and E. Lepore, *Holism, a Shopper's Guide*.

others. It is probably the case that defining holism one way rather than another has the effect of making some particular problems related to it easier to discern and manage. If we stick to the definition in terms of a distinction between properties, then it seems that two main interesting areas of investigation come into view. The first concerns an investigation into one particular subset of properties, namely semantic properties like 'is meaningful', or 'is a symbol', or 'expresses a proposition', or 'has a referent'. Are semantic properties atomic, anatomic or holistic? The second area of investigation concerns the question whether there are any grounds to believe that a slippery slope argument will oblige us to say that, for any semantic property, if it is anatomic then is must be holistic.

Since my main concern up to now has been not to pursue any one possible line of investigation but to evaluate and compare some important definitions of holism, I will not be engaging myself in these questions. Before anything can be attempted with any working definition, one must consider at some length a problem that apparently hits holism whatever its definition.

2. The All-or-Nothing Problem

Radical holism makes understanding a simple everyday term practically impossible. To understand one word, one must somehow have access to the meaning of the entire language. The argument starts from the simple observation that to understand a word, for example 'cat', one must grasp its meaning. But the meaning of that word is not obtainable as an isolateable package. According to semantic holism, the understanding of 'cat' is dependent on the understanding of the sentences in which it is used. Hence, I understand 'cat' if I understand sentences like: 'Cats are four-legged', 'Cats have eyes', and so on. To claim to have an understanding of 'cat' is to claim to have an understanding of a reasonably large set of similar sentences. But a radical semantic holist will not know where to stop with this semantic regress. In fact, other words used in each of these sentences will require their own regress for understanding *them*. Hence to understand 'cat' I need to understand 'eyes', and to understand 'eyes' I presumably have to understand 'light', and so on. Each individual language-user will have a deficient understanding of 'cat', and each user's understanding will be different. In other words, endorsing this radical semantic holism leads to the conclusion that no one understands 'cat' as I do unless they understand all my thoughts which I accept as true or might accept as true. This is difficult to reconcile with the usual idea of how language is learned and used in communication. The way we acquire the mastery of a language seems to necessitate its breaking down into significant parts. One learns a language incrementally: one learns one sentential constituent after another. As regards theories, we say we understand a theory when we are capable of deriving its significance from its internal structure. If the entire set of beliefs has to be involved each time I consider a single belief, then there can be no internal structure of a theory which provides an incremental understanding of the entire theory. As regards communication, what lies in jeopardy is our usual understanding that linguistic and theoretical commitments of a speaker and hearer can overlap partially. With holism, one will not be entitled to claim that there is such a thing as the understanding of part of another's language without understanding the whole of it.

The particular form this problem takes in the domain of philosophy of science is the much discussed problem of incommensurability. Holism seems to oblige one to work with the assumption that the meaning of a theoretical term in a given theory is determined by the entire set of sentences within the theory containing the term. Any change in the postulates containing a given theoretical term will bring about a change in meaning of that term. Hence, it is claimed that if Einstein and Newton were to meet and discourse about mass and force, they would fail to comprehend each other. There would be no question of agreeing or disagreeing because concepts like mass and force are fully dependent on the theories they are embedded in. The two theories are incommensurable. At one time, this result concerning incommensurability was considered reasonably acceptable because of the hope that holism applies only to theoretical terms: observation terms were considered directly applicable to experience. Hence the hope was that, even though Einstein and Newton would talk past each other if they tried to talk about mass and force, they would nevertheless engage in intelligent conversation allowing for agreement and disagreement if they talk about, say, the position of the planets as seen from the earth. The all-or-nothing problem however has deeper repercussions in the philosophy of science because it eradicates even this kind of hope. Once the dichotomy between theory and observation is considered untenable, then all observation statements become theory-laden. Hence both theoretical and observation terms are fully dependent on the theory in which they occur. Einstein and Newton would not even be able to avoid incommensurability in discourse about positions of planets, movements of needles, and so on.

Such incommensurability is considered detrimental to a realist understanding of science because realists want to safeguard the idea that theories get progressively better, or get progressively closer to the truth. Hence they need some kind of commensurability between theories. An argument can be presented showing that if one is committed to safeguard some kind of commensurability between theories, so as to ensure that one can rationally choose between rival theories, then one has to avoid semantic holism. The crucial point here concerns the relation between the meaning of terms and their reference. In a most general sense, the meaning of a word, for example 'cat', is related both to those things in the world, namely cats, to which it applies and also to other words with which it combines to make sentences which can be used to make assertions, ask questions, and so on. This much is uncontroversial. However, if instead of 'cat' we have 'mass', the issue will be more complicated. In this case, the identity of the person who *uses* the term apparently makes a big difference. Suppose it is Newton who uses the term 'mass', and suppose we are semantic holists situated within an Einsteinian framework. We feel confident in claiming that for Newton, there was some aspect, or a cluster of aspects, in the world related to 'mass' just like cat is related to 'cat'. But from our standpoint, we cannot easily determine what this aspect was. To determine it we would have to understand his entire theory together with all other areas of his knowledge related to his physics.

The problem of determination arises from the fact that the major access we have to this aspect or cluster of aspects is through the relation the word 'mass' had to other terms in the theory of that time. The relevant conclusion to be drawn is that, if the meaning of a term is made dependent on something other than its relation to other terms in the theory, that is, if the meaning of a term is made dependent on something non-linguistic, then it would be easier to safeguard meaning-commensurability between theories. The less holism we have in our theory of meaning, the easier the comparison of terms. In other words, the less the meaning of a term is determined by the beliefs of the protagonists of the theory about the putative referent of that term, the greater the chance that we can regard those protagonists as referring to the same thing by the same expression.[17]

The strength of this argument will be assessed towards the end of this chapter. It has been presented here to show that holism and incommensurability seem to be two philosophical positions so mutually dependent that they float or sink together. If this point is viewed in conjunction with the previous point about language in general, the conclusion to be drawn is that the all-or-nothing problem, whether in its general form or in the particular form it takes in the philosophy of science, seems to be sufficiently serious to make the case for holism appear very weak. Language-learning, communication, and commensurability will be seriously undermined if holism is accepted.

---

17 This point is made by W. H. Newton-Smith, *The Rationality of Science*, p. 163.

I will now spend some time presenting and evaluating two interesting lines of argument that allow a possible compromise solution, the first one mainly due to Michael Dummett and the second to Davidson.

Molecularism

One can say that Dummett starts his treatment of holism and its problems at the same point as Quine.[18] They both accept that language is best viewed as an articulated structure whose sentences lie at different depth from the periphery. But Dummett wants to add something significant to this image. For him, language is not just a homogenous and evenly spread continuum of inter-related sentences, the only difference between which is the 'distance' from the periphery. He claims that language is best viewed as a *stratified* structure of inter-related sentences. Language is analogous to a multi-story building. In Dummett's view, it is the disregard of this important addition to Quine's image that leads directly to the all-or-nothing problem, especially to the enormous difficulties in giving an account of the progressive acquisition of language: 'For holism, language is not a many-storeyed structure, but, rather, a vast single-storeyed complex; its difficulties in accounting for our piecemeal acquisition of language result from the fact that it can make no sense of the idea of knowing part of a language.'[19] The analogy of the multi-storey building enables Dummett to claim that one can go beyond atomism by taking into consideration chunks of language bigger than the individual sentences. Such chunks consist of sets of thoughts which are related to one another in such a way that the meaning of each is dependent on the meaning of all the others within the set. Furthermore, the analogy enables Dummett to hold that introducing new expressions into the language or into the vocabulary of a particular speaker depends upon our first constructing the lower storeys by different means.

What Dummett suggests therefore is to trace a line mid-way between atomism and radical holism. He calls this position molecularism. He does not give a summary definition of it in any major essay, but makes a number of valuable suggestions all along his works. Some of his more relevant descriptions are the following:

---

[18] My sources for Dummett's views on molecularism are the following essays: 'What is a Theory of Meaning? (I)'; 'What is a Theory of Meaning? (II)'; 'The Philosophical Basis of Intuitionistic Logic'; 'The Justification of Deduction'. See also: N. Tennant, 'Holism, Molecularity and Truth'.

[19] M. Dummett, 'What is a Theory of Meaning? (I)', pp. 137-138.

(1) On a molecular view, there is for each sentence, a determinate fragment of the language a knowledge of which will suffice for a complete understanding of that sentence. Such a conception allows for the arrangement of sentences and expressions of the language in a partial ordering, according as the understanding of one expression is or is not dependent upon the prior understanding of another. (That it be, or approximate to being, a partial ordering, with minimal elements, seems to be required if we are to allow for progressive acquisition of a language.) On a holistic view, on the other hand, the relation of dependence is not asymmetric, and in fact obtains between any one expression and any other: there can be nothing between not knowing the language at all and knowing it completely.[20]

(2) Individual sentences carry a content which belongs to them in accordance with the way they are compounded out of their own constituents, independently of other sentences of the language not involving those constituents.[21]

(3) [E]ach sentence may be represented as having a content of its own depending only upon its internal structure, and independent of the language in which it is embedded.[22]

According to this view, then, if we are to share the thought that P, then there must be other thoughts that we also share but these other thoughts do not exhaust all our thoughts. In quote (2) above, Dummett does not specify *how* a sentence can have a content in accordance with the way it is made up of constituents. Given quote (3), one must take Dummett to be saying that each of the constituents has a definite meaning of its own, and that the meaning of the entire sentence will be the sum of these constituent meanings. But such an interpretation makes sense only if the definite meaning of each constituent is an analytic statement. Otherwise, such a 'molecule' of language will not be self-contained, and the semantic regress mentioned before will be inevitable. Hence Dummett's view is that the smallest part of language that can express the proposition that P is the part that includes all the propositions to which P is analytically connected. If we take the problem of incommensurability, Dummett's molecularism would presumably open a way of solving it by suggesting that, to understand the Newtonian concept of mass, a person within an Einsteinian framework would not need to know the entire language characteristic of the Newtonian framework but only those parts of the theory which are statements analytically connected to 'mass'.

---

[20] M. Dummett, 'What is a Theory of Meaning? (II)', p. 79.

[21] M. Dummett, *Truth and Other Enigmas*, p. 222.

[22] M. Dummett, *Ibid.*, p. 304.

But, from what has been said, this way of proceeding is vulnerable. Dummett's view of language differs from Quine's version not only because of the kind of structure language is considered analogous to, but also because it assumes the tenability of the analytic-synthetic distinction. Now, since Quine's argument against this distinction is very convincing, there seems to be no clear way how Dummett's proposal of cutting up language into chunks bigger than a single sentence can be carried out in any principled way.

Revisionism as Regards Learning

A second way of responding to the all-or-nothing problem is to be revisionist as regards our views on how language-learning takes place. It does not seem unreasonable to suggest that little children do not learn language incrementally. It does not seem unreasonable to hold that they do not learn sentence by sentence but in a way similar to light dawning over the whole. Davidson gives substance to this suggestion by bypassing the idea of learning parts of a language and employing the related, but different, idea of *partly* learning the entire language.[23] According to the received empiricist view he wants to distance himself from, we first learn a few names and predicates that apply to medium sized objects through ostension. Then we advance by acquiring complex predicates and terms for objects not yet perceived. Then come theoretical terms, often learnt by being embedded in suitable scientific discourse.

The received view, he argues, is not plausible because even at the very early stages, names and simple predicates associated with everyday life cannot be assumed to function properly in isolation, if we are not dealing with very special cases. A name or predicate functions properly when incorporated within some set of similar names and predicates. The image often used to express this is the one according to which language is 'organic'. Presumably, by this term Davidson means that language acquisition, especially in the early stages, is so intertwined with the living conditions and overall behaviour of the learner, that the model of piecemeal appropriation is very implausible.[24]

But if we have successfully passed from an explanation in terms of 'learning parts' to an explanation in terms of 'partly learning', this does not mean that radical

---

[23] D. Davidson, 'Theories of Meaning and Learnable Languages'.

[24] Quine is in agreement with this at least as regards some particular idioms. He comes to the conclusion that the general term and the demonstrative singular are, along with identity, interdependent devices that the child of our culture must master 'all in one mad scramble'. Cf. W. V. O. Quine, *Word and Object*, p. 102.

semantic holism has been vindicated. What happens in the early stages of learning may be different from what happens in the later stages. If it is plausible to abandon piecemeal acquisition in the very early stages, this does not mean that such piecemeal acquisition should be abandoned for *all* stages of language-learning. It seems likely that, after a certain age, children do learn some things incrementally. Piecemeal acquisition and holistic acquisition are probably intertwined ways of acquisition co-existing throughout much of a person's life. Davidson's attempt to give substance to the idea of 'partly learning' a language does not seem to be robust enough to justify a revision of the received view. And this shows that the kind of radical holism discussed so far cannot be well defended from the all-or-nothing problem by making this revisionist move.

Let me recapitulate briefly. The overall impression is that the case for holism looks rather weak. In the first section of the chapter, three ways of defining holism were discussed and briefly evaluated: confirmation holism, semantic holism and attribute holism. In this section, the major problem that they all have to face has been described and the prospects of two lines of response were considered. The first response consisted in molecularism. The all-or-nothing problem would be solved if a mid-way option between atomism and radical holism is plausible. But this position is dependent on analyticity, and given the convincing arguments against the analytic-synthetic distinction, it seems unlikely that adherents of molecularism will be capable of giving a clear idea of the boundary defining the chunks of language which act as semantic units. The second response consisted in a certain amount of revisionism as regards language acquisition. Hence the all-or-nothing problem would be solved if the idea of piecemeal acquisition is abandoned in favour of the idea of some kind of 'organic' view of language. But this response leaves a lot to be desired as regards precision: although we may use the metaphor of light dawning over the whole, one cannot apparently be very clear about what this means. Moreover, although the organic view looks plausible for the early years of childhood, one cannot disregard the fact that piecemeal acquisition does occur, especially in later learning. To give an account of the dividing line between holistic acquisition and piecemeal acquisition seems a very difficult task.

Given these difficulties, it is probable therefore that the accounts of holism mentioned in the previous section were too ambitious. A more cautious strategy is to be less ambitious and seek the minimal account of holism which can be reasonably defended without incurring the full burden of the all-or-nothing problem.

3. Minimal Cognitive Holism

Following Jane Heal, I will start by distinguishing the following four forms of holism.[25]

(H1) The presence in a set of thoughts of one with a given content imposes some constraints on the contents of the rest of the set.

(H2) Only whole languages or whole theories really have meanings, so that the meanings of smaller units are merely derivative.

(H3) The meaning which an individual thought has depends upon the whole collection in which it occurs in such a way that any change in the whole collection (the addition or removal of an element or the substitution of one element for another) changes the content of every thought in it.

(H4) We cannot make sense of there being just one thought.

These theses cover much, if not all, of what the previous discussion showed holism to be about. The versions H2 and H3 allude directly not only to *one* thought but also to a *set* of thoughts. Endorsing any of them will thus bring up the question which has already assailed us, namely the question whether the entire set is involved or just a subset. Therefore, since the all-or-nothing problem will certainly arise if we endorse H2, or H3, it is reasonable to start by refraining from trying to justify any of these. The main aim of the rest of the chapter is to see whether the minimal version H4 is defensible. The argument in favour of H4 will include H1. H4 is here being considered a minimal version of holism because it does not contain any claims neither about how thoughts are linked together nor about how many thoughts are being considered related to any given thought. To show that this version is plausible I will elaborate an argument sketched by Heal.

Her starting point is the claim that a subject can be considered a unified locus of cognitive virtues. A subject has an ability to perceive veridically, to remember accurately, to reason validly and so on. It is unified in the sense that the cognitive states which the subject arrives at by the exercise of one ability are available to affect appropriately the manner in which the other abilities are exercised. We conduct our life with the assumption that we are subjects in this sense.

To assess whether this starting point is plausible enough, one should recall that the notion of subject has had a long history. Attempts have been made to explore what sort of thing the subject is, if indeed it is a thing at all. In traditional terms, a distinction may be drawn between substantival and non-substantival theories of the

---

[25] J. Heal, 'Semantic Holism: still a Good Buy'. I am indebted also to her *Fact and Meaning*.

subject or the self. The former contend that the subject is a substance, physical or non-physical, and the latter that it is a mode of a substance. Saying that the subject is a mode of substance seems to undermine the importance of the concept, and hence also the legitimacy of using it as a plausible premise for any kind of argument. Hence, the Humean contention that the subject, or self, is nothing but a bundle of different perceptions, is a possible worry for someone like Heal who wants to justify H4 by employing the notion of subject. Likewise, it could also be worrying to recall that some philosophers distrust our impulse to imply the existence of a subject from the way each one of us says 'I think'. According to this objection, one needs to escape the control that grammar has on our way of thinking. We escape this control by making the effort to say '*It* thinks' rather than '*I* think'. In this way we do not have to assume a self or a subject any more than we do when we say 'It's raining'.

These reflections should certainly make us careful in our handling of the concept of subject, and could perhaps also make us try to do without it. But some other cogent reasons push in the opposite direction and suggest that it would be unwise to give it up completely. Giving it up completely would mean giving up the idea of permanency through change of perceptions and ideas. Having such an invariant hinge is an essential support for the very idea of movement and change, and also for the very idea of a distinction between speaker and speech. So, it seems that the idea of subject does play a very significant and fundamental explanatory role in our understanding. Accepting this fact is enough to show that Heal's starting point, that there is such a subject, enjoys considerable plausibility.

We can safely proceed therefore with her argument in favour of H4. She makes a number of points in response to the approach taken by Fodor and Lepore. The more important of these for my purposes are the following three: first, the claim H1: that the presence in a set of thoughts of one with a given content imposes some constraints on the contents of the rest of the set; second, that the content of a thought is dependent on the context; third, that a creature with just one isolated thought is impossible. I will take each of these points in turn.

It is useful first to clarify some technicalities. The nature of some mental states, like pains, seems to be completely exhausted by what it feels like to have them. These mental states are not 'about' anything. Other mental states, like believing that p, are about things, for example about snow being white, about the cat being on the mat, and so on. These latter states, called propositional attitudes, have therefore a content, which is typically specified by the 'that-clause': I believe *that p*. More generally, 'content' is often taken to be a term for whatever it is a representation has that makes it semantically evaluable. Thus a statement is sometimes said to have a proposition or truth condition as its content; a term is sometimes said to have a concept as its content.

Now, the first point made by Heal is that the presence in a set of thoughts of one with a given content imposes some constraints on the contents of the rest of the set. To see why this should be so, suppose a subject has a train of thoughts with contents c_1, c_2, ..., c_n. Each of these contents cannot be completely independent of the others, because if the set of all c_i's contains too many contradictions, the thinker having the train of thoughts cannot be attributed with a reasonable degree of unity, and with a reasonable ability to reason. Hence the thinker will not be a subject as described above. Admittedly, a subject may hold bizarre views, but these views must be excusable on pain of undermining the original requirement that the thinking subject is a unified locus of cognitive virtues. Hence some constraints must be present within the set. What this point is opposing therefore is the view that thoughts can sit next to each another in a thinker's head without any constraints whatsoever, just like items in a supermarket ready to be picked up in accordance to some purpose or recipe.

The second point Heal makes is that the content of a thought is dependent on the context. She makes use of the observation that many other attributes are also context-relative. Consider the attribution of flatness to a field. One may rightly say that a field is flat if one is thinking of landing a helicopter, but one would wrongly say that the same field is flat if one were thinking of playing croquet on it. What makes one attribution right and the other wrong is the relation of the spatial configuration of the field to the purpose one has in mind.

Is there any reason to believe that content-attribution is like flatness-attribution? An example of how content-attribution is context-dependent will supply these reasons. Suppose you know a fair amount of chemistry and I know none.[26] The thought 'This is water' attributed to you will be related to many other beliefs which I do not have. Your beliefs about the external substance with which your thought is correlated will be different from mine because you have beliefs about its chemical composition and I do not. Your thought therefore is 'richer' than mine and therefore different. In fact, on this view, it is unlikely that any two persons will have the same thought about anything since it is unlikely that they have the same beliefs associated with the terms which express that thought. But now let us look at the content of these thoughts. Not all the aggregate of beliefs that an agent associates with 'water' are relevant to a particular context involving an action

---

[26] The example is taken from A. Bilgrami, *Belief and Meaning, the Unity and Locality of Mental Content*, p. 11.

within a particular situation. If you and I are both thinking of drinking some substance from the kitchen tap because we want to quench our thirst, our thoughts may in this locality both be attributed the same content: 'that water will quench thirst'. In this kind of context of explanation, therefore, your chemical beliefs are simply not among the beliefs selected from your idiolect. Although our idiolects are never likely to be the same for any single concept, in many contexts we may nevertheless share many contents.

We can also consider examples involving the same person. If I say 'This is water' while sitting with friends at table in a restaurant, the content of the thought is not the same as when I say 'This is water' while finishing a process of chemical analysis in a laboratory where the breakdown of the cellulose molecule is being investigated. In the first case, the content can be described by the statement: 'This is water as opposed to an alcoholic drink', while in the second case it can be described by the statement: 'This is water as opposed to some other liquid which is not composed of H_2O molecules'. The context shows that the thought has a different content, and a contrastive statement brings this out.[27]

So there are good reasons to hold that content-attribution is context-dependent. One concludes therefore that it is misleading to think of a thought as having one fixed content for all contexts. It is misleading to engage in a philosophical inquiry about thought and content with the assumption that thoughts are like small building blocks, or 'molecules' of mental activity whose contents can be discovered once and for all in the same way as the molecule of water was discovered to contain two atoms of hydrogen and one of oxygen.

The third and final point Heal makes is that a creature with just one isolated thought is impossible. To arrive at this conclusion, one should start with clarifying what having a concept means. For a creature to have a concept is for it to have a

---

[27] This point was inspired by work in contrastive-explanation in terms of why-questions. Cf. B. C. van Fraassen, *The Scientific Image*, pp. 126-129; P. Lipton, 'Contrastive Explanation'. Moreover, it should be noted here that the general point about context-dependence does not concern the related but different problem about whether the referent has a major role in individuating a thought. This latter problem has been discussed by imagining a situation involving earth and twin-earth. On twin-earth the substance that plays the part of water is not H_2O but XYZ, which has all the phenomenological properties of H_2O. If we say that what individuates a thought involves the referent, then my Doppelgänger on twin-earth does not have the same thoughts that I do when reflecting on the properties of water. If we hold that what individuates a thought does *not* involve the referent, then we would also hold that my Doppelgänger has the same thoughts as I do. See H. Putnam, 'The meaning of meaning', in *Mind, Language and Reality*. This debate does not apply directly to the point I want to make, because I am not concerned with the referent.

certain disposition to get into a given state under certain circumstances. Such a disposition, or ability, is built upon some, perhaps structural, features which the creature has at that time. Heal proposes that such foundational structures or features should be considered dependent on a 'segment of lived life, involving movement and interactions with the environment and so on'. The merits of this proposal will be evaluated later on.

Now suppose we are observing a creature as it interacts with its surroundings, and one particular behaviour pattern we distinguish makes us say that the creature is actually thinking: that it had a thought. According to the laws of the world as we know them, the behaviour we notice is based on a biological structure that allows it. This structure is the result of millions of years of interaction with the environment. Moreover, there is every reason to believe that such a biological structure does not allow just this particular behaviour pattern but other similar behaviour patterns as well. This implies that there must be moments when the creature manifests other ways of interacting with the world which also deserve our calling them moments of thinking. It does not make sense to attribute just one isolated thought to a creature. Heal makes this point in the following words:

> If the laws of our world are in force then the instantaneous appearance of highly structured creatures is impossible. The only way of getting such a creature, one with a structure intricate enough to support the dispositions required for thought, is by slow growth. Thus by the time there is enough complexity of potential interaction with the world for attribution of one particular thought to be comfortable, whatever behaviour supports that attribution will occur in an ongoing stream of varied interaction with the world which will surely also ground other thought attributions.[28]

Some counter-examples may spring to mind if we take the liberty of breaking the laws of the actual world we live in. Hence perhaps we can envisage a possible world where structures allowing thought do not need millions of years of interaction with the environment and where creatures or machines have dispositions supported in such a way that allows them to have just one thought. And this apparently undermines the justification of H4.

But arguments based on possible worlds are not always plausible. Possible worlds can be described as thought-situations described by counterfactual conditionals. If the thought-situation goes against many of the laws which hold in the actual world, then we often say that the possible world is 'distant' from the

---

[28] J. Heal, 'Semantic Holism: still a Good Buy', p. 338.

actual world. Some arguments based on possible worlds are vulnerable not only because the possible worlds employed are distant but also because the meaning of the counterfactual conditional describing them is indeterminate.

My claim is that the above suggestion apparently undermining H4 is unconvincing because it is weak on both fronts: it involves a possible world that is both distant and indeterminate. It will be useful, for the sake of clarity, to elaborate this point a little further. Consider three examples.

(1) Knowing that cats do not disappear into thin air, and that my cat went into the garage (which has no windows), then I know that if I go into the garage I will find it there. I am here employing some simple laws of nature. Even if I do not own any cats, the above inference may be made without the slightest problem. In everyday life we are accustomed to consider such counterfactual conditionals as true or false.

(2) In a similar way, we may employ our theories to conclude for example that if the fundamental constants in the early universe were different from the values we know, then no life would have appeared on Earth.

(3) This type of extrapolation in everyday and in scientific discourse may be carried out in other ways. For example, I may perhaps hold that 'If I were the same person as Aristotle, then I would not like karaoke'.

In the cat example, the antecedent of the counterfactual 'if the cat goes into the garage' is a thought-situation, or possible world, which is not very different from the actual one I usually am in.[29] Hence, the same laws of nature are assumed to apply. In the fundamental-constants example, the antecedent 'if the fundamental constants had a value different from the current values' is a thought-situation which is more distant from our actual one. But still we feel confident to say that the laws of nature apply in the new situation as well.[30] This is because the antecedents in both cases are determinate. Their meaning is clear and they are well defined in the context of the relevant laws of nature. In the third case, however, the one involving

---

[29] For my purposes here, I am using 'thought-situation' and 'possible world' interchangeably.

[30] Strictly speaking, it is better to say that, when fundamental constants change, what remains invariant is the form of the law rather than the law itself. The form of Newton's Law of Universal Gravitation is $F = G\dfrac{m_1 m_2}{r^2}$. This form may be instantiated by different laws according to different values of G. Hence, assuming discrete values of this gravitational constant, for the one form, we can envisage the following set of laws:
$$\left\{ F_1 = G_1\frac{m_1 m_2}{r^2}, F_2 = G_2\frac{m_1 m_2}{r^2}, F_3 = G_3\frac{m_1 m_2}{r^2}, ..., F_n = G_n\frac{m_1 m_2}{r^2} \right\}.$$

me and Aristotle being the same person, we are introducing a thought-situation which is not only very far from our actual one. The example is also one whose antecedent is indeterminate. Given the laws of nature as we know them, it seems that we face great difficulty when we try to see what being the same person as Aristotle might mean.

So the differences between the three examples can be brought out in terms of two attributes related to the antecedent, namely: (1) the *distance* of the new thought-situation, and (2) the *determinacy* of the meaning of the antecedent.[31] We have therefore four possible combinations: distant and determinate, non-distant and determinate, distant and non-determinate, non-distant and non-determinate. I will not consider the last case. My suspicion is that one can never have a possible world which is not distant from the actual one and at the same time described by an antecedent which is not determinate. The other three combinations however correspond to the three examples above. The cat-example involves a thought-situation which is non-distant, and its antecedent is fully determinate. The example about fundamental constants introduces a thought-situation which is distant from the actual world we live in, but at least the antecedent is fully determinate. The example concerning me and Aristotle is problematic on both fronts: it is distant and also non-determinate.

Returning to our previous line of argument, we can now evaluate what kind of thought-situation is involved in the suggestion that there could be creatures with a structure that allows them to have just one isolated thought. It turns out that this thought situation is as problematic as the third example. It introduces not only a distant possible world but also an indeterminate one. This can be seen from the fact that if I say: 'suppose there are creatures with a structure that allows them to have just one isolated thought,' it seems very difficult, if not impossible, to see what independent variables I will have to keep constant and what independent variables I have to change. Our very idea of creature will be in jeopardy, because it is itself linked to some laws of nature. Now, one can plausibly argue, as has already been done, that even in the case of a simple counterfactual, like 'if we strike the match, it will light', we cannot claim to know all the variables and laws involved.[32] The counterfactual I am analysing here is an extreme case of this. It is similar to the counterfactual about me being the same person as Aristotle. The new thought-

---

[31] I will simplify matters by not allowing these attributes to have degrees. I will not discuss here thought-situations that are slightly more distant or slightly less distant than others. I will not discuss antecedents that are slightly more determinate or slightly less determinate than others.

[32] J. Aronson, *A Realist Philosophy of Science*, pp. 237-259.

situation goes against so many of our normal thought constraints that it seems very difficult to admit that such phrases have any meaning even though they seem to have a *prima facie* meaning.

So one can safely conclude that the objection to Heal's third point depends on possible worlds that are so remote that the objection becomes ineffective. This concludes the argument in favour of H4. The three conclusions were: first, that the presence in a set of thoughts of one with a given content imposes some constraints on the contents of the rest of the set; second, that the content of a thought is dependent on the context; and third, that a creature with just one isolated thought is impossible. These three points, taken together, give sufficient justification for the claim that we cannot make sense of there being just one isolated thought.

Conclusion

The overall line of argument has been the following. In the first section, some kinds of definitions of holism were discussed and evaluated. In the second section, the all-or-nothing problem was described. Two attempts at defending holism from this problem, namely molecularism and revisionism, were briefly presented but it was shown that they both face formidable difficulties: the first because the viability of the analytic-synthetic distinction is not acceptable given Quine's convincing arguments, and the second because the notion of 'partly learning' a language is problematic. The option was then taken to seek a justification only of a minimal version of holism according to which one cannot conceive of just one isolated thought.

Three short remarks are in order. First, what is minimal about this version of holism? Saying that one cannot conceive of just one isolated thought does not include anything about how advocates of H4 are to solve the all-or-nothing problem. Although the justification supporting H4 is convincing, important questions are being left unanswered, the most pressing of which seems to be the following: if we accept minimal holism as described in H4 and also want to avoid the all-or-nothing problem, aren't we then obliged to opt for some kind of molecularism? An affirmative answer here seems likely. But we are still faced with the daunting task of deciding how large the 'molecules' must be, and whether they should have definite boundaries. H4 does not contain any direct information about the criterion that should determine which thoughts agglomerate with which. There are therefore a fair number of reasons why H4 is being called minimal.

The second remark concerns a question which comes naturally as a consequence of the first: if H4 is so minimal, is it worth holding on to? One may have the impression that a vast battery of arguments has been marshalled for a claim with a very modest content. One may therefore be tempted to think that very little has been achieved. Wouldn't it have been better to start just with the unproblematic claim that there is in fact more than one thought? The answer is no. The non-trivial content of the minimal cognitive holism defended here can be appreciated much better if two of its more important features are spelt out.

First, when one holds that we cannot make sense of there being just one isolated thought, one is endorsing the view that the set of thoughts of any individual thinker cannot be considered a simple aggregate, a mere collection of individual and separable thoughts. The aggregate in this case possesses an internal relationship between its parts such that, as the familiar dictum says, the whole is greater than the sum of its parts. The main point therefore is about the nature of this aggregate of thoughts we normally call mind. In general, one can see that aggregates are of different kinds. One way of differentiating between them is to determine a scale of varying degrees of holism.[33] On one extreme we have aggregates like a heap of sand where the relation between part and whole is minimal. As we move towards the other extreme, the relation between the whole and its proper parts becomes more and more significant, until we have aggregates where a part cannot exist apart from the whole.[34] The cognitive holism endorsed in this chapter signifies that the mind is indeed an aggregate with a very high degree of holism. The proper parts of the mind, including thoughts, beliefs, and other propositional attitudes, are related to each other and to the whole they constitute in such a way that, for each part to be what it is, it must be part of a mind. For a thought to be a thought, it must be part of the aggregate.

The second important feature of the minimal version H4 is that its justification is not dependent on the problems discussed in the earlier sections of this chapter. Even though many related issues remain unresolved, the justification of H4 described in the preceding text is not directly related to these unresolved issues. H4 is justified on other grounds. Hence, not knowing how to deal with the all-or-

---

[33] Given the complexity of the world, allowing degrees of holism is better than considering only two kinds, namely holistic and non-holistic systems, as is done by M. Esfeld, 'Holism and Analytic Philosophy'.

[34] This happens when the relation between part and whole is constitutive of each. As will be discussed in Chapter 3, a good example of a holistic aggregate is a community of persons. A person, because of the requirements of rationality, can only receive full realisation from the community of which he or she forms part.

nothing problem, and not having enough criteria to decide how large the 'molecules' ought to be, should not make us abandon holism in this minimal version. It is the nature of cognition that requires us to accept H4. A consequence of this is particularly relevant to the issue of incommensurability already alluded to. As we have seen, Newton-Smith's approach to the issue of holism involved taking holism and commensurability as opposing positions. According to him, if the first wins the other loses, and vice versa. To defend realism, and hence commensurability between theories, his option was to endorse a non-holistic semantic theory for scientific terms. This option is certainly understandable because, as we have seen, if one accepts radical semantic holism, the problem of incommensurability will be unavoidable. His move however should not be taken to mean that all kinds of holism are detrimental to the same extent to a realist understanding of science which allows commensurability between theories. Now, I have concluded that cognitive holism of a minimal kind is justified on independent grounds. Hence, my project in the following chapters will be similar to Newton-Smith's in seeking a plausible realist account of science, but unlike him I will be giving cognitive holism a central role to play in the articulation of that account.

The third and final remark concerns an important consequence of the line of argument justifying minimal holism. It was said that a creature has a concept, or believes something, when it has a disposition to get into a certain state under certain conditions. Furthermore, what makes possible such an attribution, namely the attribution 'has a concept' or 'believes that p', is a structure which depends on the biological makeup of the creature and also on a segment of its life, and which allows that particular disposition. As already mentioned, Heal suggests that the structure depends on a 'segment of lived life, involving movement and interactions with the environment and so on'. What this suggestion amounts to, I propose, is the view according to which the structure can be thought of being affected in two distinct ways. It can be affected by long term interaction with the environment, interaction that affects not only the one creature under consideration but the entire species it belongs to. And it can be affected by short term interaction with the environment, interaction that affects the one particular creature. When we are dealing with human thinkers, the second aspect implies that the structure allowing dispositions depends on the particular person's interaction with his or her environment, an interaction which involves non-linguistic conventions, practices, rituals and so on.

The line of argument justifying H4 therefore opens up the interesting — but difficult — possibility of considering being a symbol, or being a meaningful part of language, as internally related to playing a role in a system of non-linguistic conventions, practices and rituals. Taking a step further in this direction will make one have doubts whether there is any strong justification for an assumption I have been indirectly making all through this chapter, namely the assumption that language is a set of sentences. Such an assumption starts to seem too narrow. What is more plausible now is the claim that linguistic behaviour is continuous with non-linguistic behaviour, and perhaps even with culture in general.[35]

Hence, although the cognitive holism discussed so far is minimal in some respects, it does nevertheless open up a number of interesting routes for philosophical analysis. These routes seem particularly appropriate for my overall project of examining how the understanding of science can be undertaken without avoiding the complexity arising from the interaction between its theoretical and practical aspects.

This kind of cognitive holism, however, does not allude directly to the fact that science is a group activity. Attention in this chapter was not centred on the fact that situations often involve a multiplicity of inquirers. Cognitive holism deals primarily with what happens in one isolated mind. The next move therefore has to be a search for another complementary aspect of holism covering aspects other than what happens in one isolated mind.

---

[35] Linking together linguistic-behaviour and the entire domain of gestures, expressions, actions, practices, rituals, and ways of life is called by Fodor and Lepore anthropological holism, but it is not dealt with in any detail. See J. Fodor and E. Lepore, *Holism, a Shopper's Guide*, pp. 6-7, 209.

Chapter Three

Social Holism

Social holism is the position according to which one cannot properly conceive of just one isolated inquirer. Many debates in philosophy have been centred on situations involving one apparently isolated inquirer. Hence, for example, Cartesian epistemology has often been presented as concerned with questions like 'how do I know I am really awake?' or 'could my belief that I am sitting by the fire turn out to be false, since I might be asleep in bed and having a dream?'. Such kinds of debate are concerned mostly, sometimes exclusively, with the cognitive processes of a single inquirer. However, one can easily observe that such questions used by the doubting Cartesian make sense only within a meaning-generating community of inquirers. So, although philosophical investigation into the cognitive processes of one isolated inquirer may not be completely irrelevant to the understanding of science, we usually have some preliminary intuitions that highlight the relevance of a philosophical study that emphasises the relation between inquirers and its impact on cognition.

Social holism and cognitive holism are complementary notions. As explained in the previous chapter, according to cognitive holism, one cannot conceive of just one isolated thought. In other words, the individual mind is an aggregate with a very high degree of holism. Thoughts, beliefs, and other propositional attitudes, are related to each other and to the whole they constitute in such a way that, for each of these to be what it is, it must be part of a mind. When discussing the structural features of a creature allowing the dispositions needed for its thoughts, it was pointed out that the growth of these structural features involves interaction of the creature with its environment. What will occupy my attention in this chapter, put briefly, is the claim that this interaction involves not only the effects on the creature of its *physical* surroundings, which presumably shape the mechanism of cognition in the long run according to the laws of evolution, but also the interaction between the creature and other similar creatures. Again, one can realise that there is intuitive support for this claim, especially if the creatures are not considered in the abstract but taken to be human beings engaged in science. Few would deny that any account of science will be distorted if no mention is made of some interaction between scientists.

But intuitive justification is not enough. In the text that follows, my main aim is to formulate an argument justifying social holism in a way that is more cogent than these intuitive suggestions. In summary form, the argument runs as follows. The claim that one cannot properly conceive of one isolated inquirer will be understood as equivalent to the more precise claim that to evaluate, discuss or even understand the beliefs of one inquirer, one has to consider many other inquirers as well. The crucial element in the argument to illustrate this is the *expression* of the beliefs of the inquirer under consideration. I am assuming that the expression of beliefs must contain words. Hence I will not spend time speculating about the possibility of an inquirer having beliefs that cannot be put into words. If such beliefs are at all possible, I am proceeding on the assumption that having such beliefs is not relevant for my concerns. A first plausible way of arguing for social holism will be to show that the sense of the words used by an inquirer to express his or her beliefs must bring in the entire community of inquirers in which those words are used successfully. In the first section, this line of argument will be discussed in some detail, but it will be shown to lack sufficient generality. For a more general argument, the expression of a belief is taken to involve the intentional attempt on the part of the inquirer to shape his or her thoughts with a view to having them satisfy conditions of rationality. Another line of argument will therefore be followed in the second section of the chapter to show that the conceiving of one inquirer's beliefs, understood in this more complex way, also necessitates the conceiving of that inquirer as being together with others. The third and final section will consist in defending social holism from possible problems arising when talking of group belief.

1. Using Words

The meaning of words is socially dependent. This claim, if plausible, allows a direct way of justifying social holism because it will imply that, for an inquirer to grasp the sense of the words he uses to express his beliefs, he must take into consideration an entire community of inquirers. An argument for the social character of meaning can be put forward in the following way.

The reference of an expression is the entity it stands for. Sometimes, two different expressions have the same referent. Hence 'the morning star' refers to the same thing as 'the evening star', and yet the sentence 'the morning star is the evening star' is informative while the sentence 'the morning star is the morning star' is not. To solve this puzzle of identity, Gottlob Frege introduced the notion of sense to be understood as a mode of presentation of reference. Hence I may present an entity according to *my* knowledge that it is observable mainly in the mornings, while another inquirer may present the same entity according to *his* or *her* knowledge that it is observable in the evenings. According to this understanding, the sense of a given expression is primarily defined with respect to the knowledge of one particular inquirer. Suppose we ask: what is *the* sense of some given expression? Then, on the view sketched so far, which is that of Frege according to most commentators, the only answer available is that *the* sense which an expression has is that sense which all, or most, *individual* inquirers attach to it. Because sense is determined by individual inquirers, the only option available is to account for *the* sense by equating it to the result of a kind of averaging procedure. One can realise from these features that this view goes against the claim of social holism: it implies that single inquirers can be properly conceived on their own because every individual has all the required baggage for the determination of sense.

This simple view however is wrong. It cannot handle simple cases where the full sense of a word is unknown to the one who uses it. Hilary Putnam argues in the following way. One inquirer may use the word 'elm' even though he does not know how to tell an elm from a beech, but knows only that it is used for some species of trees. Such an inquirer intends that anything he says about elms should be judged true or false by reference to other speakers of the language, especially those entitled to claim authority in the matter, like botanists and lexicographers. What makes it useful for such an inquirer to use the word 'elm' rather than just 'tree', even though he does not know the difference, is that he may accept some statements about elms on the testimony of others.

These observations may tempt us to think that, contrary to the Fregean view, the sense of a given expression has nothing to do with the knowledge of the one who uses the expression. It has to do only with the knowledge of experts. If we say this however we would be effectively going to the other extreme position, which is also in general unacceptable. The knowledge belonging to the user of an expression, in most cases, does have a central role in determining the sense. The case discussed by Putnam is the exception not the rule. Hence, what seems to be more plausible is to seek a middle way between Fregean individually-determined sense and the extreme position according to which sense is determined exclusively by experts. Dummett in fact suggests the following way of correcting the Fregean individualism.[1] One may hold that in many cases there is a sense which is attached to a word by most speakers of a language, or by those who have authority as to its

---

[1] M. Dummett, 'The Social Character of Meaning'.

use, which is that sense to which a speaker of the language will normally intend to be held responsible. This way, we can account for the elm case discussed above: a speaker may attach only a partially specific sense to the word 'elm' but may still make himself dependent on the existence of the socially accepted sense to enable him to use it in communication.

After these considerations, we can appreciate how being aware of 'division of linguistic labour' helps us to articulate the way one inquirer is essentially dependent on others. A situation wherein one inquirer has beliefs about gold, for example, should be understood as follows: since that individual is a member of a community, what is accessible to that individual is a *partial* sense of 'gold'. If the individual is a professional chemist, then he or she will probably have a more sophisticated sense of 'gold'. But the position portrayed so far should not be taken to imply that only chemists grasp the full sense of the word 'gold'. That would be counter-intuitive: normal speakers under normal conditions can legitimately claim that they understand well enough what gold is. The meaning of the word is fully conveyed neither by a description of criteria employed by the experts nor by a description of those used by ordinary speakers. It involves both. It involves also a grasp of the relationship between them. This grasp of the relation between them consists of further social criteria that determine which individuals or groups are to be acknowledged as genuine experts. This view has the plausible consequence that the sense of a word is not something static, fixed for all time, but could change according to what ordinary speakers do in ordinary circumstances. What counts as ordinary in one period of history is not the same as what counts as ordinary for other periods of history. Moreover, the sense of the word changes because the knowledge of experts is constantly changing as well, and also because the social criteria determining who the experts are may change as well.

To recapitulate, therefore, a first option was considered according to which the sense of a word is understood as the way of presenting a referent. This presentation is carried out in line with the knowledge of the user of the word. This option could not handle the many cases where users do not grasp the full meaning of the words they use. Hence a better option is to hold that the sense of a word is a complex attribute that involves at least the three elements: first, the knowledge of the user, second, the knowledge of the experts associated with the area of knowledge linked to that word, and third, the criteria determining which people are to be acknowledged as experts.

This argument certainly gives credibility to the thesis of social holism. But its generality is dubious. It is meant to show that meaning of words is socially dependent. But what was shown in fact was that meaning is socially dependent as far as *certain* words are concerned: like 'gold', 'beech', 'elm'. One can make a tentative extrapolation from this argument and suggest that meaning is socially dependent for *all* other words, even those whose meaning does not require experts of any kind, like 'and', 'or', 'this', and similar simple words. But such generalisation needs more argument to make it convincing.

2. Following Rules

I will therefore begin with the point that beliefs are often expressed as sentences constituted by many words. Such sentences could contain words which certainly involve some division of linguistic labour. They could contain words which certainly do not. And they could contain many words which lie somewhere in between: words about which we cannot be certain whether their meaning depends on experts or not. Whatever words are used in the expression of a belief, we can safely say that the expression involves the intentional attempt on the part of the inquirer to shape his or her thoughts with a view to having them satisfy conditions of rationality.

Some remarks should be made here about the distinction I am drawing between beliefs and thoughts. This distinction was not needed up to now. What was said before in this chapter and also in the previous chapter, applies both to thoughts and to beliefs. Here however the distinction starts becoming significant. I will be taking a belief to be a mental state with a proposition as its content. I will follow the usual way of considering belief as the primary cognitive state: other cognitive states, like perception and knowledge, are to be considered as a combination of beliefs and other factors. Some beliefs are at the forefront of consciousness. They are things one is, at the moment, thinking about and feeling to be true. But not all beliefs need to be at the forefront of consciousness in this way. Even if you thought about it once, when you were at school, you were not consciously thinking, a moment ago, that whales are not fish. However, it is correct to say that you believed it even when you were not actively thinking about it. Moreover, even if you *never* consciously thought about whether foxes play the clarinet, it seems right to say that you believed that they do not play the clarinet, before your attention was ever drawn to the fact. On the grounds of these considerations, I will be taking a thought to be whatever is presently, or at a specified moment, occupying my attention. I will be taking a belief to be a thought, possible or actual, which I feel is true. To *believe* that *p* is not the same as to *accept* that *p*. Belief has to do with feeling, while acceptance with reasoning. Hence, I will be taking a belief that *p* to be a disposition to spontaneously feel that *p* is true. To accept that *p* is to treat *p* as a premiss in my reasoning, whatever my spontaneous feeling about it.[2] Since acceptance that *p* very often causes the belief that *p*, I will be assuming that beliefs worth considering are those that are, or can be, accepted as premisses for reasoning. This is why my starting point is the observation that the expression of a belief involves the intentional attempt on the part of the inquirer to shape his or her thoughts with a view to having them satisfy conditions of rationality. If the inquirer discovers that some of his beliefs are contradictory with others, then he will be worried and will make the necessary adjustments by accepting some and rejecting others.[3]

The next step is to observe that one who has beliefs and expresses them is a thinker who must be able to treat something as a rule, so that he or she will be able to determine when it is appropriate to accept various beliefs. But to be able to treat something as a rule entails some other obligations. A subject can treat something as a rule only if some conditions are satisfied regarding the nature of rules. The most intriguing condition is that the rule in question must be a normative constraint over an *indefinite* variety of cases, so that one can determine the correct response in each of those cases. Hence, to investigate the expression of accepted beliefs in terms of the intention of the inquirers to be rational, one must struggle with the well-known problem of rule-following: how can something that is normative over an indefinite number of cases be suitably accessible to a finite mind?

Consider a very simple rule, associated with the concept of 'box'. I use 'This is a box' to identify boxes. This simple rule is something that dictates a certain response in an indefinite range of cases, and it is also something that I can identify and try, at least, to respect in all these cases. Notice the difference from the previous argument for social holism. The centre of attention there was the sense of a given expression. It becomes evident that the sense is different from the referent when two different expressions refer to the same thing. In the more complicated cases considered here, the centre of attention will be the working of a typical concept, like that of 'box', which is essentially one expression with many possible referents. From the case of various expressions with one referent I am now moving to the case of one expression with various referents. The concept of 'box' will be considered as a simple rule telling the user which response is acceptable. One cannot consider this rule as something fixed in the mind, as a kind of 'quale', as an

---

[2] Cf. L. J. Cohen, *An Essay on Belief and Acceptance*, chapter 1.

[3] There is also the case, which will be discussed later on in this chapter (§3), of an inquirer accepting that *p* but not believing that *p*. This case involves a certain detachment on the part of the inquirer. As far as rationality is concerned, such an inquirer is also, and perhaps even more, intent on having what he accepts satisfy conditions of rationality.

image or formula. The rule associated with the concept of 'box' cannot be reduced to a set of paradigm boxes because, as will be shown, it would be possible to treat such paradigm boxes as instances of other rules.

Saul Kripke has famously described this by emphasising that, if we start with a finite number of typical cases of addition, say, then nothing can guarantee that novel cases should follow suit. Any rule identifiable from the finite number of cases is underdetermined. Other rules may be consistent with the finite number of cases and yet diverge at some future point from the usual addition rule we are familiar with. This may be illustrated by the plus-function and the quus-function. A finite set of examples of addition, like $1 + 2 = 3$, $2 + 2 = 4$, $3 + 1 = 4$, cannot determine the right way when novel cases are encountered. The plus-function as normally understood forces us to say that $68 + 57 = 125$. But the examples cannot distinguish this from the quus-function, where this quus-function is the same as normal addition with the exception that when the numbers to be added are greater than 57, the result of the process is 5. 'Since the sceptic who supposes that I meant quus cannot be answered, there is no fact about me that distinguishes between my meaning plus and my meaning quus. Indeed, there is no fact about me that distinguishes between my meaning a definite function by "plus" (which determines my responses in new cases) and my meaning nothing at all.'[4]

So any set of paradigm boxes can be extrapolated in an infinite number of ways. A serious problem therefore must be faced by those who make the legitimate claim that a rule must be a normative constraint over an *indefinite* variety of cases, and yet be accessible to the kinds of finite minds we have.

A convincing solution to the problem has been put forward by Philip Pettit.[5] To give a plausible account of what is happening when I use the simple rule 'This is a box', one is allowed to say that a finite set of examples can *exemplify* a rule even though, as Kripke has shown, a finite set of examples cannot *instantiate* a rule. Hence, I start with a set of paradigm boxes which exemplify the box-rule for me. I am naturally inclined to go on from these paradigm boxes by associating other boxes which do not differ too much, maybe only in colour or size. I will not

---

[4] S. Kripke, *Wittgenstein on Rules and Private Language*, p. 21. Other relevant texts consulted for the question of rule-following are: P. Boghossian, 'The Rule-Following Considerations'; G. P. Baker and P. M. S. Hacker, *Wittgenstein: Rules, Grammar and Necessity*; S. Holtzman and C. Leich, (eds.), *Wittgenstein To Follow a Rule*; P. Pettit, 'The Reality of Rule-Following'; T. Thornton, *Judgement, Objectivity and Practice*; L. Wittgenstein, *Philosophical Investigations*; *Remarks on the Foundations of Mathematics*; C. Wright, (ed.), *Special Issue on Rule-Following*.

[5] P. Pettit, *The Common Mind*, pp. 76-106.

associate chairs, or dogs with the paradigm cases, because the difference is considerable. Admittedly, there are many border-line cases. The paradigm boxes fix the rule for me because I depend on an extrapolative inclination. This inclination is trustworthy when the situation I am in can be described as normal. It is not a completely arbitrary inclination. It is constrained by at least two important factors. First, there is constancy over time within me as an individual. I would be worried if I am aware that my inclination leads me in different directions at different times. I would be worried if I am aware that my inclination makes me call a milk carton a box at one time, and it makes me not call it a box at another time. Second, there is also constancy in the decisions taken by other followers of the rule. Hence, I would be worried if I am aware that my inclination leads me in a different direction from that followed by others as regards what a box is. I would be worried if my inclination makes me call a milk carton a box while my friend's inclination is not to call it a box at all. These two factors may be taken to constitute what the expression 'normal circumstances' means. Pettit explains this in the following words:

> [...] there are two assumptions that we spontaneously and systematically make as participants in an area of discourse when we form and discuss our beliefs. These are assumptions, respectively, of intrapersonal and interpersonal constancy. The intrapersonal assumption is that something is amiss if I find myself reliably inclined to make different judgments at different times — in particular, judgments different by my own lights — without any justifying difference in collateral beliefs or whatever. The interpersonal assumption is that something is amiss if you and I find that we are reliably inclined to make different judgments — again, judgments different by our lights — without any such justifying difference.[6]

Hence, the account of rule-following developed so far allows the following procedure: first a subject takes certain cases to exemplify a particular rule; second, the subject reads off the rule in new cases by letting the inclination lead her on as to how she should go; third, the subject allows that negotiation with herself over time, and negotiation with others, may establish that the circumstances are not favourable, and that therefore the rule was mistakenly read.

One should notice at this point that without such interaction between the inquirer at one time and the same inquirer at an earlier time, or an interaction between the inquirer and other inquirers, rule-following cannot be satisfactory.

---

[6] P. Pettit, *The Common Mind*, p. 93.

Fallibility will not be possible, and I am taking fallibility to be one of the necessary constituents of rational thought.

Now, it has been argued in the Wittgensteinian literature connected with the private-language argument that, although we have some strong reasons to believe that rule-following involves being corrected by others when necessary, we cannot exclude the possibility of a socially isolated rule-follower, who interacts consistently with herself to secure the fallibility necessary for rational thought.[7] In other words, we cannot exclude the possibility that the negotiation I may engage in with my past self is enough to secure some minimal kind of genuine rule-following procedure. As opposed to this, others have argued that grasp of a rule must be manifest in what is interpersonally accessible in a such a way that there is no such thing as a kind of rule-following which is intrinsically unknowable by another. At least, there must be the logical possibility for a person to grasp the rule being followed by another.[8]

I will not get involved in this debate. The context within which I am arguing for social holism allows me to follow a more modest line of argument suggested by Pettit.[9] I say 'more modest' because a resolution of the above debate in favour of the second position described will constitute a transcendental proof that it is *in principle* impossible to have beliefs by employing concepts available without interaction with other people. This is a very strong thesis. Since the social holism I am advocating will be applied to the understanding of the scientific enterprise, I am entitled to bring into these considerations one important aspect associated with this particular enterprise. The rules followed in science are rarely, if ever, rules *meant* to be followed by one person in isolation. There may be cases where a prominent scientist will presumably innovate ways of thinking or of experimenting that involve rules which, for a certain duration at a certain period in history, are known only by himself or herself. However, in these cases, it seems very plausible to assume that the new rules are always *knowable* by others. The rules which are of relevance to the understanding of science must have this special property. They must be knowable by others. This is not to say that they must be known. In fact, there is nothing claimed about what the others know or do not know — there is nothing about whether there are any others around or not. The important point is that rules that are of relevance to the understanding of science are such that they are knowable by others. To say that a rule is knowable is to hold two things: first, that

---

[7] S. Blackburn, 'The Individual strikes back'.

[8] C. Wright, 'A Cogent Argument against Private Language?'.

[9] P. Pettit, *The Common Mind*, pp. 175-193.

at least one other person may reliably identify the rule, as opposed to engaging in pure guessing; second, that this other person may follow the same rule. If there are such things as rules which may properly be followed and yet which are in principle unknowable by others — and I am not committing myself one way or another on this point — then my assumption here commits me to hold that these rules will not be relevant to the understanding of science. This is plausible enough because, if say it is discovered in some archive that Fermat had proved his last theorem by following rules that are in principle unknowable by others, few would consider the theorem a respectable result. The knowability of the rules is what we value.

It may be helpful here to mention Pettit's way of expressing this point. He makes use of a metaphor. 'Commonable land is land that is or may be held in common by people, where being held in common is a condition that requires awareness of the sort of joint control exercised. By analogy, we might say that according to our publicity condition the rules that people individually follow, in particular the rules that they individually follow in their thought, are commonable ... They are rules over which no one individual has a monopoly; they are capable of being claimed as a common possession by any of the individual's fellows.'[10] My suggestion here is that, if one accepts that commonability is plausible as regards rules of thought in general, then one can accept that commonability is even more plausible as regards rules of the kind of thought that is relevant to the scientific enterprise.

The general point I am interested in can now be expressed by the following question: if one inquirer follows rules that are knowable by others, is interaction with himself or herself at different times enough to ensure proper rule-following? The answer is no. The inquirer needs interaction with others. If this can be shown, then social holism follows.

Suppose that the way I try to follow a given rule is fixed exclusively by reference to my own responses. Certain paradigm examples induce in me an inclination to go on in a certain way. I trust my inclination except when I find that, according to my own introspection, it leads me in different directions at different times. In this situation, another inquirer can only make *guesses* as to what rule I am trying to follow. The observer will conjecture that it is this rule or that rule: what is available for her is the set of observable choices I actually make. The observer will never be in a position to check her hypothesis by negotiating with me. She knows that I am negotiating with myself over time. But for all she knows, I may not be negotiating with myself over time in the same way as she would. She could keep on

---

[10] P. Pettit, *The Common Mind*, p. 180.

conjecturing about the rule I am following. She could even falsify some of her conjectures when she finds that I do not respond as she had predicted. But the problem is now apparent. It is the same problem discussed by Kripke: when an observer sees me following a rule that is fixed solely by reference to my own internal interaction, then she is in the impossible position of someone who is trying to identify, from a finite number of cases, a rule that is relevant for an infinity of cases. The conclusion to be drawn therefore is: if I follow rules only through interaction with myself over time, the rules will not be knowable in any strong sense. Others can only guess.

I will now take the other corresponding scenario: I picture myself trying to follow rules by heeding not only what discrepancies may appear in the interaction with myself at earlier times but also what discrepancies may appear in the interaction with my observer. In this case, there may be convergence of my responses and her responses. In the negotiation with my observer, the responses I give according to *my* extrapolative inclination may converge on the responses she would give according to *her* extrapolative inclination, or they may not. If convergence occurs, then the rule I was following is something knowable by my observer. It is knowable both in the sense that she can identify it rather than guess it blindly, and also in the sense that she can follow it herself. In this scenario then, I do not have an absolute jurisdiction over what rule I intend to follow. If I am honest, I will intend to pick out a rule which is identified not only through interaction with myself but also through how my observer responds under favourable conditions.

It may happen obviously that no convergence occurs between my responses and the responses of the one observing me. It may happen, in other words, that, given the paradigm examples to start with, I find my inclination leading me to extrapolate in one direction while the inclination of my observer will lead her repeatedly to extrapolate in another direction. This scenario has a simple explanation. It can safely be taken to mean that there is no target to our intention. The rule I thought I was following does not exist. Pettit puts this in the following words: 'If there is no such convergence, then your project is empty in the way in which it would be empty to hunt the snark or try to search out the present king of France; it fails to come off, because it lacks a target.'[11]

So, in summary form, the argument for social holism, as developed by Pettit, starts from the claim that to have beliefs, a given inquirer must follow rules. Adding the requirement that these rules are knowable by others, we are obliged to hold that

---

[11] P. Pettit, *The Common Mind*, p. 188.

the inquirer has to check his rule-following by reference to what others are inclined to do in the same domain. If an inquirer is following rules fixed only by reference to his own responses, then any observer cannot but observe him 'from the outside', as it were. This situation entails the impossible task of identifying a rule on the basis of a finite number of cases. As opposed to this, if an inquirer is following rules fixed both by reference to his own responses and by reference to those of observers around him, then the observer is 'on the inside'. This ensures that the rules are knowable. The upshot is that to express accepted beliefs, an inquirer must follow rules of rationality, and, to follow rules, an inquirer must be in interaction with others. Hence it is not possible to conceive of just one isolated inquirer: others must be brought into the picture.

It is essential at this point to acknowledge that, although the argument is plausible, it is founded on a simple, idealised picture of rule-following. I will mention two interesting ways how Pettit's approach may be developed to cover more ground, especially as regards the understanding of science.

First, interaction with others is not always straightforward. If we are dealing with the rule associated with 'is a box', the situation is quite accessible. Many people, if not all, would roughly agree as to what counts as a box and what does not. But if we are dealing with the rule associated with 'is desirable', the situation will be more difficult to handle. Admittedly, if the simple cases show that an inquirer always follows rules fixed both by reference to his own responses and by reference to those of observers around him, then the complex cases probably should follow suit. However, in the complex cases, one question starts becoming more pressing: who are these others? If I am the one doing the rule-following and if the observers I am interacting with are 'of my type', then extrapolative inclinations coincide easily. These inclinations coincide more easily than if the observers were 'of another type'. For some rules, the group of people I interact with makes a difference. Admittedly, one may hold that some rules are real in the sense that they can be targeted by any two inquirers of whatever type. The rule of addition is presumably one of this type. One may hold also that some other rules are not real in this sense: they can be targeted only by two inquirers who have some things in common, for example they can be targeted by two scientific inquirers only if these two belong to the same Kuhnian paradigm. But there are apparently no cogent reasons to believe that a neat, distinct boundary divides these two types of rules. It is very probable that the reality of rules as described here comes in degrees. And this is something Pettit has not included in his account of rule-following but which seems essential to highlight when scientific activity in all its complexity is the object of study.

Second, there may be a significant temporal gap between checking my following a rule by reference to my own previous responses and checking it by reference to the responses of observers around me. The gap will be particularly significant in historical studies. In such studies, scholars sometimes attempt to determine the kind of rules that some prominent scientist was following in a certain period of history. But the details regarding interaction with others may not be available. In such cases, obviously, we are back to guessing — we are back to all the Kripkean insecurity which that brings with it. This point highlights the fact that Pettit's account may be called synchronic: the rule-following and the interaction with others are happening during the same time period. But one often has to deal with a situation which is diachronic: the rule-following and the interaction with others, which is essential to ensure that the rules are knowable, do not happen during the same time-period. These cases allow for the possibility of a person being somehow ahead of his or her time. One person's introspection may be of a kind that we can plausibly conceive of him or her targeting rules which, although knowable, are perceptible to others only after a certain learning experience on their part.

So this refinement on Pettit's account allows the following situation. Suppose one prominent scientist is being looked upon with concern because the rules she is following seem to be inaccessible. Colleagues may legitimately ask: 'since we cannot follow the same rule as she seems to be following, is it the case that she is not targeting any rule, or is it the case that she is targeting a rule which we have not yet learnt how to perceive and follow?'

This kind of question is important for two main reasons. First, it is relevant to our understanding of science because the entire approach it represents allows plausible accounts of some important cases in history, like the one involving the clash between Galileo's way of arguing, both about the world and perhaps more so about the interpretation of the Bible, and the way of arguing in line with established rules followed by the intelligentsia of the time. The question is important not only because it intuitively corresponds to what happens in such actual cases, but also because it moreover offers a possibility of understanding expertise. An expert is essentially, form our point of view, someone whom we credit with having the skill of following rules which to us are difficult or impossible to follow but which we hold are in principle knowable, at least by some, either at that time or at some later period. The second reason why the question is important is because it enables us to understand better some opaque concepts introduced by philosophers of science to account for scientific revolutions, such as Thomas Kuhn's notion of conversion.[12]

---

12 T. Kuhn, 'Postscript', in his *The Structure of Scientific Revolutions*, pp. 174-210.

Protagonists of an old scientific theory often start using a new theory by translating it into the vocabulary of the old. Such use of the new theory is mediated by the bridge of translation. Protagonists who start this way may later on experience a transition, which Kuhn calls a conversion, whereby they will not need the translation procedure anymore but will use the new theory as if it becomes their own native language. In the approach sketched so far, a conversion experience may be understood as the act of perceiving and starting to follow a rule which had not been followed before. In the case of conversion, the responses one group gives according to the group's extrapolative inclination will converge on the responses the other group gives according to its extrapolative inclination.

If these two points of refinement are taken into consideration, the justification of social holism becomes more convincing than the argument we started with. Now, one is entitled to hold that meaning is socially dependent not only for words whose sense requires recourse to experts but also for all kinds of words. But before concluding the discussion here and moving on to see how social and cognitive holism affect our understanding of science, I want to examine whether accepting social holism will lead to insurmountable problems. What accepting social holism leads to is, roughly stated, a shift of attention from what happens within one isolated inquirer to what happens within groups of inquirers, as was mentioned at the very beginning of this chapter. I will therefore start the following section by clarifying the notion of group, and then proceed by solving a problem that arises when trying to conceive of group beliefs.

3. Group Belief

Groups are collective objects described and analysed in the social sciences.[13] They represent in fact just one such collective object. In the most general sense, it is useful to distinguish between things and events. I am taking things to be particular entities situated in space and time; they exist for more than one moment but have no temporal parts: like tables, mountains and so on. I am taking events to be temporally extended; they have temporal parts which are also events: like illnesses, chemical reactions, and so on. When considering social entities, we can have the same distinction. There are social things, like families, tribes, and scientific

---

[13] I am indebted here to D-H. Ruben, *The Metaphysics of the Social World*; D. H. Mellor, 'The Reduction of Society'.

communities. There are also social events, like wars, elections, famines. Groups are social things of a certain kind.

If we assume groups to be abstract objects, then they will also be causally impotent. But this assumption is based on a confusion between a group of people and set of people. Sets are abstract objects but groups are not the same as sets. There are usually two sets with which a group may wrongly be identified, namely the set of its present members, and the set of all its members past, present, and future. In both cases, there are problems. Sets are defined by their members. Different members means different sets. Groups, however, are not like that. A tribe does not change into another one when one member dies. Groups, to a certain extent, survive a change in their members. To say that groups change even if the number of its members changes by one is to fall into implausibility at one extreme. To say that groups are sets of all their members, past, present, and future, is to fall into implausibility at the other extreme: we would not be able to conceive of a group changing over time at all. Moreover, two distinct groups can have precisely the same members but sets cannot. For example the Library Committee can also be a famous dancing group: both groups can have the same members. So groups are not equivalent to sets.

To determine a set of individuals, it is enough to specify some property. For example all blue-eyed people in a certain College constitute a set: as a set, it is abstract and causally impotent. If other criteria are added, then it may become a group. Groups exist only when individuals with some common property are regarded, or regard themselves, in certain socially significant ways, whereas sets of individuals exist quite apart from such socially significant further considerations. If I single out the blue-eyed individuals in College, I have singled out a set in terms of a common property. For this set to constitute a group, either of two conditions must be satisfied. Either the one doing the singling out, namely myself in this case, has to start thinking of them as people with some social purpose, for example a group which needs to receive special optical treatment, or the group which is often discriminated against while travelling in a given country. Or the individuals themselves may band together into a group by having some jointly accepted view. It is plausible to think of scientists in a laboratory, for example, as forming a group because, apart from having the common property of being in the same location for a considerable portion of the day, which is enough only to constitute a set, they all have the added socially significant property of sharing some specific jointly pursued purposes and some specific jointly accepted views. D. H. Mellor has suggested that a group can be defined by some 'founding fathers' just as an individual human being may be said to get his or her identity from a particular

DNA configuration fixed at conception. But the purpose that brought the 'founding fathers' together is very important because it allows a plausible account of how groups continue over time even though they do not have the selfsame parts as they had to start with. The continuity in carrying out this purpose fixes what at a later time constitutes the same group. Steady replacement of the parts is possible. People may leave or replace others in a group, just as cells do in a body without affecting its identity, so long as enough remain at any time to keep the whole thing going. This point about the importance of a purpose in the definition of a group should be noted, because it will be the corner stone of the account of practice I will be offering in the next chapter.

Given these preliminary reflections about groups, I am now in a position to tackle the problem that arises with group belief. If we try to conceive of beliefs as possessed by a group of inquirers, rather than conceiving them as possessed by just one isolated inquirer, we will apparently be obliged to accept either of two alternatives. Either we accept a summative account of group belicf, according to which for a group to have a certain belief most members of the group, taken as individuals, have that belief, or we accept that social groups have a mind of their own.

Both alternatives lead to unacceptable situations. The first alternative, which involves the acceptance of a summative account of group belief, taken in its simplest version, does not say anything about whether each of the members of the group should or should not know that the others have the belief in question. If we say that a group believes that p, are we saying that most of them believe that p, or that most of them believe that p and they know it? To refine this account further therefore, we can add the corresponding condition by holding that a group believes that p when all or most members believe that p and this is common knowledge in the group. Common knowledge here can be understood roughly as the idea that the fact that the members all, or nearly all, have the belief that p is itself known and believed by all, or nearly all, of them and they realise this.[14] This refined version is however still unsatisfactory. It does not account for some common situations where one member of a group, or one small sub-group, carries the others along in their accepted beliefs simply because of authority or prestige. In such cases, the group can legitimately be said to have one belief, namely the one put forward by the leader and accepted by the others as a group belief, even though most of the members taken individually have another belief, and everybody knows it. I am assuming that such situations occur also in scientific activity. Groups of scientists

---

[14] D. Lewis, *Convention*.

do not always have beliefs that can be understood in the summative view. Groups of scientists can be influenced by the authority and special skill of certain members. The summative account therefore lacks the intricacy needed to be effective when describing the cases I am interested in. Limiting my analysis to it is therefore an unsatisfactory option.

The other alternative does not fare any better. A belief, in so far as it is a propositional attitude, can be the possession only of entities that are capable of having attitudes, namely minds. So, to avoid the summative account and still claim that groups have beliefs, we will be obliged to hold that groups have minds, or collective consciences. This option is attractive because it allows us to talk of the group having a belief even if all, or nearly all, its members are said not to have this belief. But to allow this, we must ascribe to the group a self-conscious mind and mental feelings of its own which it has independently of its members. This ascription, apart from being intuitively doubtful, entails the heavy burden of suggesting and justifying some philosophical theory concerning the ontological requirements for such minds to exit and function. On this option, therefore, social holism, although minimal in its formulation, seems to entail a burden which few philosophers would be ready to countenance.

In the remainder of this chapter, I will be suggesting a way out of this dilemma. I will start by recalling the distinction made earlier in this chapter between belief and acceptance. There I said that a belief that p is a disposition to feel it true that p. Acceptance that p is to treat it as a premiss that p. Hence belief has to do with feeling, while acceptance with reasoning. So to avoid the limitations of the summative account in the first option, and avoid also the heavy ontological requirements of the second option, I will make a first suggestion as follows. When we endorse social holism and find ourselves accepting statements of the form 'a group believes that p', we are not talking strictly speaking of belief, but essentially of acceptance. Hence we are not dealing with feeling p to be true, but of treating p as a premiss by the members of the group for decisions taken as a group.

Group belief has been analysed on these lines by Margaret Gilbert, and it is useful to mention some of her conclusions.[15] She starts by considering how groups of people negotiate to arrive at a joint view. She offers the following example to illustrate what such negotiation involves. A group of people meet regularly to discuss poetry. A poem is read and each participant is free to make suggestions as to its interpretation. In this situation, Gilbert sees that what is both logically necessary and logically sufficient for the truth of the ascription of group belief is

---

15 M. Gilbert, 'Modelling Collective Belief'; *Social Facts*.

that all or most of the members of the group have expressed willingness to let a certain view 'stand' as the view of the group. Once a group view has been ascertained, say by the statement 'in the opinion of the group, the last line is quite moving', then subsequent remarks which go against this view should include some kind of excuse: 'I know that we thought that the last line is moving, but personally I have my doubts'. A certain *obligation* is taken on by members of the group when and in so far as a group view is established.

Here, joint acceptance of a view has to be distinguished from an individual acceptance and also from the majority acceptance of a view. A view I *individually* accept is my own personally accepted view. But a view I accept as the jointly accepted view of the group I am in is not necessarily the view I personally accept. I may be willing to jointly accept that *p* in spite of my personal acceptance that ~*p*. For an example of this, one can refer to the history of quantum mechanics during the first half of this century. In spite of the opposition to Niels Bohr's views by some leading scientists like Albert Einstein and Erwin Schrödinger, the vast majority of physicists accepted the complementarity interpretation in general without reservations, at least during the first two decades after its first appearance. That this Copenhagen interpretation became the jointly accepted view during this period can be concluded from the fact that most text-books written between 1930 and 1950, including books by Alfred Landé, Louis de Broglie and David Bohm, who later all turned against the Copenhagen view, advocated the complementarity principle according to which the numerical value of a physical quantity has no meaning whatsoever until an observation upon it is performed.[16] The fact that the personal accepted beliefs of some individual scientists were different did not show in the jointly accepted view.

Similarly, one must realise also that when most members of a group each accept a certain view, it does not necessarily follow that that view will be the jointly accepted one of that group. One may argue that justice is done only if the majority view is identical to the jointly accepted view. But justice as understood here is not always the determining factor in the emergence of jointly accepted views. Members of a group may not feel bound to accept the majority view. They may want to follow a charismatic leader, someone with authority, an expert. In fact, the majority view before the formulation of the jointly accepted view is often different from the majority view after its formulation. The process of negotiating a jointly accepted view may affect the personal beliefs of members of the group, and therefore it may affect what they will accept as premisses.

---

[16] M. Jammer, *The Philosophy of Quantum Mechanics*, pp. 247-251.

Using this notion of a jointly accepted view, Gilbert suggests the following definition of group belief:

S   A group G believes that *p* if and only if the members of G jointly accept that *p*.[17]

This concise statement may be broken up into finer vocabulary by saying that a group G believes that *p* if and only if the members of G have openly and intentionally expressed their willingness to accept that *p* together with the other members of G as a body.

Now although, this definition of group belief points in the right direction, in so far as it escapes the dilemma mentioned earlier, it needs serious revising to be applicable with a reasonable degree of success to actual cases encountered in science. I will mention two main areas where revision is essential.

The first concerns the interaction between belief and acceptance. Statement S does not emphasise the difference between these two propositional attitudes. It therefore neglects to a certain extent the importance of the beliefs of members of a group. If we subscribe to what was said previously, namely that a belief that *p* is a disposition to feel it true that *p*, and that acceptance that *p* is to treat it as a premiss that *p*, then the temptation is to suggest that scientists are people who accept the theories they propose without being emotionally involved. We may suggest, in other words, that scientists accept theories without believing them. In this way, it is hoped that scientists will be free from the danger that arises when the possession of a belief might make them less ready to change their mind about accepting the contrary if new evidence becomes available. In this picture, scientists are people who practice complete intellectual detachment: they do not have any beliefs that may influence their acceptance of theories either personally or jointly in a group.[18]

---

[17] M. Gilbert, *Social Facts*, p. 306.

[18] Belief and acceptance of scientific theories have been discussed in a different way by B.C. van Fraassen, *The Scientific Image*. According to the distinctions I am working with, his treatment of the issue is based on an implausible presupposition. In his view, realism is the position according to which acceptance of a scientific theory involves the belief that it is true, while his antirealism is the position according to which acceptance of a theory involves the belief that it is empirically adequate (p. 8; p. 12). Both positions are described in terms of acceptance, and in both positions this acceptance involves belief. This seems to mean that such acceptance is *caused* by a certain belief. If one recalls that a belief is essentially a feeling, one sees that his distinction presupposes that scientists and philosophers, whether realists or antirealists, are led by their feelings. Hence his view is the exact opposite of the naive view presented here according to which scientists always exhibit complete detachment

But this is only an unreachable ideal. In actual practice, joint acceptance of views in science are intricately intertwined not only with beliefs in the actual theories under consideration but also with beliefs about the competence or expertise of individual scientists. Consider as a historical example the relatively recent episode concerning John von Neumann's proof that quantum mechanics is logically incompatible with hidden variables, published in 1932. It was thirty-four years later that John Stuart Bell published the first explicit identification of the mistake it involved. It is arguable that von Neumann's prestige as a mathematician greatly influenced other physicists who accepted his view as the jointly accepted one, even though the Copenhagen interpretation was still considered as a kind of new constraint not fully justifiable. John von Neumann's method of doing physics, namely the method of axiomatisation, derived from the recognised authority, David Hilbert, can be said to have enhanced the belief in his contemporaries that he is an unquestionable expert. At least one physicist writing recently still holds that 'the authority of von Neumann's overgeneralised claim for nearly two decades stifled any progress in the search for hidden-variables theory. This is especially surprising because of the obviousness of *inapplicability* of one of von Neumann's axioms to any realistic hidden-variables theory of the kind that attempts to explain quantum theory as a special case'.[19] This same physicist continues: 'So, what we have been discussing should have been obvious. The truth, however, happens to be that for decades nobody spoke up against von Neumann's arguments, and that his conclusions were quoted by some as the gospel. There must be some magic in his arguments that could fool people into believing that *his* definition of hidden-variables theory would be the only correct one rather than the obviously inappropriate one.'[20]

The version of group belief expressed by S therefore needs revision firstly to make the clear distinction that beliefs have to do with feelings and acceptance with reasoning, and secondly so as to include the inevitable pressures that arise due to beliefs that members of a group will have as regards the views being discussed and about the persons who propose and believe them.

---

from any feelings. Neither complete detachment from feelings nor complete dependence on feelings seem to offer the adequate account of science being sought here.

[19] F. J. Belinfante, *A Survey of Hidden Variables Theories*, p. 24.

[20] *Ibid.*, p. 34. Cf. L. Caruana, 'John von Neumann's "Impossibility Proof" in a Historical Perspective'; here I argue, among other things, that the blame cannot be put exclusively on von Neumann, as Belinfante seems to imply. In the proof, von Neumann did in fact express himself in a clearly cautious way.

The second area where statement S needs refinement when applied to a study of science has to do with the element of obligation within a group that jointly accepts a certain view.

From S, we can see that those who participate in joint acceptance of a view thereby accept an obligation to do what they can to bring it about that any joint endeavours among the members of the group be conducted on the assumption that the view is a premiss for future decisions. This obligation is not binding to the extent that, if the group jointly accepts that *p*, then each member must believe that *p*. What the obligation entails is that, if a member has a different belief, and she wants to express this different belief, then she should make it clear that she is speaking in her own name and not in the name of the group. The account of group belief advocated here says nothing about the reasons why members of a group come together and are willing to jointly accept one particular belief. Once the belief is jointly accepted, however, some sense of obligation, even perhaps intimidation or coercion, must be present.

Moreover, a group may form a judgement at a meeting from which some of its members are absent. These members may not know of the judgement at the time. In such a situation, the judgement qualifies as a group belief, according to Gilbert, only on condition that there is an understanding or convention among the group members that absentees are to be considered as obliged in the usual way by the outcome of the meeting. Hence for the judgement to count as group belief, those who are absent are to be considered as obliged to jointly accept the judgement. The absentees must be assumed to have said: 'whatever the group comes up with on this issue, I will jointly accept with all the rest.'

Now the problem that arises when such considerations are applied to scientific activity is the following. There is a hidden assumption that the idea of a jointly accepted view is usable on condition that the group involved has strict boundaries. Statement S presumes that group decisions are made only in situations similar to a sitting of a committee, where each member can accept the view openly and willingly. When one is examining scientific activity, such a restricting assumption is certainly implausible. The locus of a scientific process leading to a jointly accepted view is nearly never a committee meeting, as may be imagined, say, for decisions taken by political parties, or clubs.[21]

---

21 From the history of science, one can identify some rare occasions which approach this kind of decision-making: for example, international meetings for the standardisation of units, or meetings meant to impose political restrictions on scientists, like what happened in Russia in 1948, when D. Lysenko delivered a speech to silence all opposition by announcing that his own 'report' had the support of the Central Committee of the

For a plausible explanation of what occurs in scientific jointly accepted views, we must first hold that the locus of a decision is an extended event without clear boundaries consisting of various aspects such as the evaluation of a received view, a conscious or unconscious decision about which questions are relevant for a given hypothesis, an experiment involving particular and perhaps novel techniques, the interpretation of experimental results, theoretical conjectures involving some thought experiments, and so on. I will call this extended event consisting of the entire complex of these various aspects an intervention. A loosely defined group of scientists can be considered to have a jointly accepted view as the outcome of the processes I am calling interventions. There will certainly be some members of the group who are involved, to some extent or other, in all the processes involved in an intervention. But most members will not be directly involved in all of the processes. For example, the experimenter may not know exactly what the statistician does with the data, and the statistician may not know exactly how the data was collected. In spite of this compartmentalisation, a tacit agreement is always present, namely that if any rational inquirer had fully been through the entire intervention, then he or she would have been willing, because *obliged* by rationality, jointly to accept the view expressed by all those involved in the intervention.

So, for the account of group belief, as stated in S, to be applicable in a satisfactory way to the understanding of science, the element of obligation assumed to exist within a group should not be seen as the imposition of a well-defined 'committee decision' but as an expression of the requirements of rationality. This obligation due to the requirements of rationality is expressed by the conditional statement 'if any rational inquirer had fully been through the entire intervention, then he or she would have jointly accepted the proposed view'.

Let me recapitulate. In this section, I have described a dilemma that faces anyone who accepts the social holism thesis. One horn of the dilemma consists in holding on to social holism and explaining group belief in terms of a summative account. This way was shown to be inadequate because it does not cater for the role of experts or of authority figures within a group. The other horn of the dilemma consists in holding on to social holism and explaining group belief in terms of a disposition or feeling of the entire group. This option was shown to have considerable problems because it obliges one to explain how groups can have a

---

Communist Party. The report mentioned how there were two opposing schools of biological thought in the USSR: one that was materialistic and soviet, and another that was idealistic, reactionary and committed to a chromosome-based theory of heredity. Lysenko associated himself with the 'ideologically correct'. See D. Joravsky, *The Lysenko Affair*, pp. 112-130. Such cases surely represent the exception and not the rule.

self-conscious mind independently of their members. As a way out of this dilemma, I suggested considering the term 'group belief' as a misnomer: one should understand what happens when groups agree to hold a certain view not in terms of belief but in terms of acceptance. Gilbert's version of group belief in terms of jointly accepted views was then evaluated and two main refinements were suggested. First, I underlined the fact that, although scientists may *aim* at being emotionally detached from what they accept, they cannot be detached completely: they cannot escape having beliefs which may prejudice their judgement concerning the relevance of new evidence. Hence scientific activity cannot be analysed in terms of joint acceptance only. Second, I suggested that talking of jointly accepted views in science should not be understood as implying the existence of a strictly defined group: a scientific view is not jointly accepted because of a committee decision but because of an obligation due to the demands of rationality.

My overall aim in this chapter has been to define, defend and elaborate social holism. The definition of social holism was structured to be minimal in the same sense as the definition of cognitive holism was minimal in chapter two. A first possible way of arguing for social holism was to show that the sense of the words used by an inquirer to express his or her beliefs must bring in the entire community of inquirers in which those words are used successfully. This line of argument, although successful to a certain extent, was shown to be insufficient because it is of dubious generality. An argument whose generality was more secure was then presented. It involved the investigation of the intentional attempt on the part of inquirers to shape their thoughts with a view to having them satisfy conditions of rationality. Hence section two contained an argument for social holism based on rule-following. Given that to have relevant beliefs and to express them, an inquirer must follow rules which are knowable by others, we are obliged to hold that the inquirer has to check her rule-following by reference not only to her own responses but also to what others are inclined to do in the same domain. Hence, to properly conceive of one inquirer, other inquirers must be brought into the picture. In the third section, I argued that endorsing social holism does not lead to insurmountable problems which would emerge if an unrefined notion of group belief is used. The problems are avoided if for groups, instead of the notion of belief, one uses the notion of acceptance.

I will be assuming henceforth that the claims put forward in this chapter and those put forward in the last chapter give enough reasons for the claim that cognitive and social holism should always be taken into consideration in any account of science.[22] One should recall here, for the sake of clarity, that these two kinds of holism do not merely involve the weak claim that, in this world of ours, there isn't just one inquirer but many, and that there isn't just one thought but many. What is involved is the stronger claim that there is, in this world, an inevitable interaction between the many inquirers and an inevitable interaction between the many thoughts. What is being refuted is thus the assumption that, as far as philosophical analysis is concerned, inquirers and thoughts are conveniently isolatable.

---

[22] I am not claiming that all possible forms of holism should fall under these two categories. In the literature, there are forms of holism other than the ones I alluded to, and some are specifically relevant to the way scientists deal with theory and data. For example, Nancy Cartwright has made suggestions that holism may be attributable to laws. In parallel with my cognitive holistic claim that one thought is linked to others, Cartwright's suggestion is that a physical law derived from one shielded environment is linked to other laws derived from other shielded environments: 'if the events we study are locked together and changes depend on the total structure rather than the arrangement of the pieces, we are likely to be very mistaken by looking at small chunks of special cases.' N. Cartwright, 'Fundamentalism vs. the Patchwork of Laws', p. 287; see also N. Cartwright, 'Can Wholism reconcile the Inaccuracy of Theory with the Accuracy of Prediction?'. I will not be considering this kind of holism because it deals essentially with how scientists can allow for holism in the work they do, rather than with how one should go about understanding science itself, which is my primary concern.

Chapter Four

Discourse and Practice

In the previous two chapters I have argued that a plausible kind of holism can be understood in terms of its cognitive and social aspects. The thesis that we cannot conceive of one concept in total isolation from other concepts suggests that a distortion in our understanding of science will occur if we limit our investigations to apparently isolatable concepts: if we limit our investigations to objects of inquiry like isolated theoretical terms or isolated observation statements. To avoid this kind of distortion, we are obliged to look for ways of undertaking our investigation by concentrating on larger units of language, units of language that allow for the fact that thoughts and concepts do not form a mere collection but an aggregate with a high degree of holism. The notion I will be working with will be that of discourse. This notion will be a key concept employed specifically to allow for the possibility of holism, and because of this, as described earlier, I will be referring to it as a holistic concept.

More precisely, I will say that a concept C is holistic with respect to concept K in the case when (1) C is used as a central concept in the understanding of K; and (2) C guarantees, to a certain extent, that our understanding of K will take holism into consideration. Here, understanding a concept means, roughly speaking, giving an account of it in such a way that illustrates how it is related to other concepts; the more concepts it is related to the better the understanding. Suppose, for example, we want to understand jealousy. We are bound to take some concepts as central in our process of understanding it. Suppose we employ the concept of 'brain-state' as a central concept in our process of understanding. According to the description above, the concept of 'brain-state', although central, cannot be considered a holistic concept with respect to jealousy because it includes very little direct reference, if it does include any at all, to holism. It is, in other words, bringing in the assumption that to understand jealousy one can start by concentrating on certain kinds of brain-state and then summing up so as to arrive at the understanding of the whole. The resulting understanding of jealousy will probably be distorted. On the contrary, a central concept like 'joint acceptance' or 'convention' will be, in this context, an appropriate holistic concept. Distortion can obviously still arise. But at least it will not arise as a result of avoiding holism. As far as science is concerned, the central concept of 'discourse' is holistic, but a central concept like 'theoretical term' will probably eliminate any direct reference to holism in the subsequent understanding of science; and thus cannot be called a holistic concept.

The same kind of reasoning holds for the social aspect of holism. The thesis that, in the kind of world we live in, we cannot have one inquirer in total isolation from other inquirers suggests that a distortion will occur if we limit our investigations of the scientific enterprise to what happens in the brain of an isolated inquirer, like an isolated Cartesian scientist deciding whether he knows that he is sitting with a paper in his hand. To avoid this kind of distortion, we are obliged to undertake our investigation by considering collections of inquirers and by considering collections of actions of inquirers. Such collections are often called practices. With respect to science, the central concept of 'practice' is holistic, but a central concept like 'individual, isolated observer' will probably eliminate any direct reference to holism in the subsequent understanding of science; and thus cannot be called a holistic concept.

Since the concepts of discourse and practice will be used extensively in the development of the overall argument, it is imperative that a careful analysis of each is given some attention. This is the main aim of this chapter.

1. The Notion of Discourse

If discourses are things we can intelligibly talk about, then it seems that they must have elements which are common to them all. The temptation is therefore to elaborate the notion of discourse by enumerating the most important such features common to all token discourses. It is argued by those with essentialist tendencies that only this procedure can qualify as an acceptable elaboration of the concept of discourse. Nominalists will take the opposite attitude. They will claim that discourses have nothing in common except that they are called discourses.

My way of escaping the intricacies within such a conflict consists of employing Ludwig Wittgenstein's idea of family resemblance.[1] If one were to look at different types of games, one does not see something that is common to all, but similarities overlapping and criss-crossing. Games form a family not in virtue of features common to all. Similarly, I will take discourses to form a family not in virtue of features common to all.

---

[1] L. Wittgenstein, *Philosophical Investigations*, §§ 66, 67. See also R. Bambrough, 'Universals and Family Resemblances'.

Discourse and Language-game

I will be using discourse and language-game interchangeably. I am taking a language-game, as introduced by Wittgenstein, to be a language, invented or natural, seen as analogous to a game, especially in being a non-sharply defined concept explained by reference to overlapping central cases linked in various complex ways with large numbers of different cases. Invented language-games are complete, but natural language-games are not. I am not saying that no features of discourses may be helpful in teaching someone what a discourse is. When a child is learning what a game is, the teacher may give some examples. Later on, the child acquires the capacity for correctly applying the word. Similarly for kinds of discourse: we may think of the discourse associated with the word 'pain', the discourse associated with selling cheese by using a balance to determine the weight, the discourse associated with evaluating comedy, the discourse surrounding the complex assessment of a new scientific theory, and so on. Interaction between natural language-games can be an important object of reflection: for example, one can come to realise that we talk of electric current because of the penetration of the vocabulary of hydrodynamics into our understanding of electricity in virtue of the analogy between the behaviour of fluids and electricity.[2]

Since I am working with the basic concept of family-resemblances rather than with a strict concept of universals, I do not want to commit myself to the claim that there are some attributes necessarily common to all discourses. There is one attribute however, to be found in most discourses, which is particularly significant for my interests, given my commitment to holism. The attribute is the possibility of disagreement between the protagonists of the discourse.

To see why it is these discourses exhibiting this specific attribute that are of interest to me, it is enough to recall the content of the two types of holism: cognitive and social. According to the former, we cannot properly conceive of just one isolated thought; according to the latter, we cannot properly conceive of just one isolated thinker. If the discourses I consider are those whose protagonists enjoy total agreement among themselves, then the situation will be practically indistinguishable from the situation involving one single inquirer. I will not be involving social holism. To guarantee the inclusion of this latter kind of holism, I have to ensure that the discourses under consideration include a multiplicity of points of view: and hence the possibility of disagreement between the protagonists.

---

[2] For a complete invented language-game see *Philosophical Investigations* § 2; the language-game with the word 'pain' is mentioned in *ibid.*, § 300; a list of language-games is given in *ibid.*, § 23.

The main idea to be investigated in the following pages is related to the following fundamental question: if disagreement is possible within a discourse, what can give rise to it? It will be seen that not all disagreements arise in the same way. Hence, a reply to this general question will be the main starting point for a useful investigation of what lies behind disagreements in scientific discourse.

Cognitive Command

A useful study of disagreement within discourse, in the most general case, has been initiated by Crispin Wright especially through his use of the concept of Cognitive Command. Since this concept will be used extensively in later arguments, I will give a quick overview of Wright's considerations and some points of further development. According to Wright,

CC   a discourse exhibits Cognitive Command if and only if 'it is a priori that differences of opinion formulated within the discourse, unless excusable as a result of vagueness in a disputed statement, or in the standards of acceptability, or variation in personal evidence thresholds, so to speak, will involve something which may properly be regarded as a cognitive shortcoming'.[3]

The cognitive shortcomings may be classified under three headings. He indicates that we may have firstly divergent input, that is, the disputants working on the basis of different information and hence guilty of ignorance or error, depending on the status of that information. Secondly we may have unsuitable conditions, resulting in inattention or distraction and so in inferential error, or oversight of data and so on. Thirdly, we may have malfunction, for example, prejudicial assessment of data, or dogmatic attitudes.[4]

What lies at the back of Wright's definition is obviously the very nature of representation. Put simply, if two appliances produce representations, then if conditions are suitable, and the appliances both function properly, they will produce divergent output if and only if presented with divergent input. Wright calls this the Representation Platitude. Cognitive Command consists in the direct application of this Platitude to the case of human inquirers: 'the nerve of the Cognitive Command constraint is a specialisation of that idea [the Representation

---

[3] C. Wright, *Truth and Objectivity*, p. 144; a previous tentative definition is found in *ibid.*, p. 92-93. 'Cognitive Command' is called 'Rational Command' in C. Wright, 'Realism, Anti-realism, Irrealism, Quasi-realism'.

[4] C. Wright, *Truth and Objectivity*, p. 93.

Platitude] to the case where the representational system is a thinking subject engaged in formation of belief.'[5]

The reader may realise here that Wright's notion of Cognitive Command is related to the discussion about rules mentioned in the last chapter. It is worthwhile situating this discussion here in the context of the views described there. I worked with the idea that, if I am a rule-follower, a set of paradigm cases fix a rule for me because I depend on an extrapolative inclination. This inclination is trustworthy, and not completely arbitrary, when the situation I am in involves two kinds of constraint. First, there is constancy over time within me as an individual. I would be worried if I am aware that my inclination leads me in different directions at different times. Second, there is also constancy in the decisions taken by other followers of the rule. Hence, I would be worried if I am aware that my inclination leads me in what is a different direction to that followed by others. Hence, it was shown that fallibility in rule-following can be secured by conceiving of rule-following as a process involving three steps: first a subject takes certain examples to exemplify a particular rule; second, the subject reads off the rule in new cases by letting the inclination speak as to how she should go; third, the subject allows that negotiation with herself over time, and negotiation with others, may establish that the circumstances are not favourable, and that therefore the rule was mistakenly read. It was argued moreover that, in the negotiation with my observer, the responses I give according to *my* extrapolative inclination may converge on the responses she would give according to *her* extrapolative inclination, or they may not. If convergence occurs, then the rule I was following is something knowable by my observer.

One significant similarity between these reflections on rule-following and the points made about the notion of Cognitive Command is that both accounts give a central role to the possible difference in the way protagonists react. In the account concerning rule-following, a sustained difference in extrapolative behaviour means that the original rule-follower was not targeting any rule whatsoever. In the account concerning Cognitive Command, a difference in opinion manifests what kind of discourse one is situated in: if the difference of opinion is regarded as involving a cognitive shortcoming, then the discourse is of one kind, namely the kind that exerts Cognitive Command. If not, the discourse is of another kind. In spite of this similarity, one cannot claim that Wright's reflections on Cognitive Command are reducible to the apparently simpler reflections about rule-following. There is at least one major difference. It is worthwhile emphasising this here so as to bring out

---

5 C. Wright, *Truth and Objectivity*, p. 146.

the particular advantages of introducing the notion of Cognitive Command in this chapter. The notion of Cognitive Command was introduced as an attribute of a discourse, and the concept of discourse is holistic. We are not dealing in this chapter with one idealised situation consisting in a rule-follower and an observer. We are dealing with a discourse which is meant to involve a whole set of interconnected rules.

So there are good reasons to believe that the notion of Cognitive Command will be very useful, and perhaps even indispensable, in the understanding of discourse. Two important questions, however, have to be faced as regards the adequacy of the statement CC. Firstly, are those who employ the concept of Cognitive Command obliged to be verificationist in their theory of meaning, or empiricist in their general philosophy? And secondly, is Cognitive Command a definite discourse attribute as opposed to a vague one with no clear boundaries? Discussing these questions will enable me to formulate an understanding of Cognitive Command which is appropriate for my needs in the overall argument.

To tackle the first question let me first define verificationism as follows: verificationism is a philosophical position according to which the conditions for a sentence or thought to be meaningful are identical to the conditions for its being verified or falsified. Verificationism is thus a thesis concerning meaning. Is Wright bringing in verificationism through the use of CC?

The answer certainly depends on what he takes representations to be. For the case of devices, it seems that Wright's argument consists in the claim that the Representation Platitude is constitutive of the very concept of representation-producing devices. Up to now, we have no direct need of any theory of meaning. But when Wright applies the case involving devices to the case involving human inquirers, then some theory of meaning must be incorporated because the notion of human representation is deeper than that of device representation. Unlike the case of photographic images, agreement and disagreement involving human representation depend on words and meaning. A context is involved. The question therefore is: does the use of the concept of Cognitive Command *when applied to human discourse* oblige one to take on board a verificationist theory of meaning?

It seems that statement CC is in need of some refinement, particularly because the status of cognitive shortcoming has to be elaborated carefully. What do people lack when they have these shortcomings? What they lack does not always have to include direct experience: especially if we recall that some cases of disagreement, involving human and thus non-omniscient inquirers, involve aspects that transcend evidence. Suppose we know that the disputants are in good condition and are functioning well. If the dispute concerns whether the moon is made of cheese, there is good reason to believe that the divergent input can be resolved. If the dispute however is about whether Julius Caesar had eggs for breakfast on the morning of his assassination, then whatever causes divergent input in the two disputants is, for all we know, evidence-transcendent. In this latter case, disagreement is obviously due to cognitive shortcoming on the part of *both* disputants.

These reflections indicate that Wright will indeed fall into the verificationist pitfall if he holds that it is meaningless to claim that there are sometimes sources of disagreement that are evidence-transcendent. Wright however does not have to say this.

He can likewise be non-committal as regards empiricism. I am taking empiricism to be the position according to which genuine differences in our claims must express the differences in our experiences. Wright will be empiricist if he limits the understanding of cognitive shortcoming to changes in experience. In my opinion, he comes very close to this in the way he applies without qualification the case of representation-producing devices to representation-producing human inquirers. But again, he does not have to do this. One may use the concept of cognitive shortcoming without making it exclusively dependent on changes in the experience of the inquires.

To safeguard a non-empirical, non-verificationist version one may reason in the following way. Suppose the empirical input for two inquirers is O, and suppose the inquirers disagree in what they are saying. If the discourse exerts Cognitive Command, Wright would have us believe that there is cognitive shortcoming. But the disagreement may be due not to cognitive shortcoming as regards O but to cognitive shortcoming as regards background theory. I am not taking background theory here to mean only the strictly scientific presuppositions of the inquirers. I am taking background theory to be much wider so as to include also the meaning of words used by the inquirers to articulate what they experience. Different inquirers will presumably nearly always have different access to background theory. There cannot be *empirical* input from background theory as understood here. Hence, in a revised version of CC, it should be made clear that an inquirer receives input O and moreover, unlike the simple devices mentioned in the Representation Platitude, has access to a subset of existing background theory. A case involving perfect ideal inquirers, defined as perfectly rational with *complete* access to *all* background theory, will be the limiting case corresponding to Wright's case of devices. Only for these ideal cases is disagreement explainable exclusively by empirical cognitive shortcoming.[6] So the answer to the first question is that, since there are significant differences between device-representation and human-representation, users of the concept of Cognitive Command do not have to be verificationist or empiricist.

The second question had to do with whether Cognitive Command is a definite discourse attribute as opposed to a vague one with no clear boundaries. Although in most places Wright assumes that comic discourse is definitely an example of a discourse not exerting Cognitive Command, I am not so sure that clear-cut statements like this one are applicable to natural discourses. It seems more appropriate to handle the concept of Cognitive Command in a way that avoids assuming that some discourses have it in full and some do not. Cognitive Command is better viewed as an attribute that comes in degrees: some discourses have it to a great extent and some have it to a lesser extent. According to the more plausible account I would like to work with, on one end of the spectrum we have discourses like scientific discourse, exerting a high degree of Cognitive Command, while at the other end we have discourses like comic discourse exerting a low degree of Cognitive Command; discourses like moral discourse presumably lie somewhere in the middle.

If we reformulate CC so as to include the conclusions derived from the above discussion of the two questions, we will arrive at the following statement:

A discourse exhibits a high degree of Cognitive Command if and only if it is a priori that differences of opinion formulated within the discourse will involve something which may properly be regarded either as a cognitive shortcoming concerning the experiential input or as a case of different access to background theory, or as a combination of both these factors.

So this concludes my analysis of the first holistic concept, namely the concept of discourse, which is of relevance to the understanding of science. Discourses were considered to form a family not in virtue of features common to all. One particular feature however is of particular importance for my aim, namely the possibility of agreement and disagreement. The notion of Cognitive Command was brought in to elaborate on this attribute of discourses. Two conclusions were then drawn. First that, since there are significant differences between device-representation and human-representation, users of the concept of Cognitive Command have to be neither verificationist nor empiricist. Second, Cognitive

---

[6] To be fair to Wright, it must be conceded that this point is somewhat vaguely included in his definition CC, especially in his qualificatory remarks involving 'the standards of acceptability, or variation in personal evidence thresholds'.

Command is better viewed as an attribute that comes in degrees: some discourses have it to a great extent and some have it to a lesser extent. We move on now to the second holistic concept.

2. The Notion of Practice

The usual understanding of the term 'practice' involves the idea of habitual action or performance. It involves the idea of a repeated exercise in an activity requiring the development of skill, as in the idiom 'practice makes perfect'. The term 'a practice' however is often used to refer to one element making up a religious, political, or scientific tradition. This latter notion is linked to a number of related concepts like paradigm, culture, *Weltanschauung*, conceptual scheme, and so on. All these concepts have been employed in a number of related disciplines in attempts to describe what happens when one particular tradition does, or does not, undergo a transformation in the course of history. As has been indicated previously, the relevance of these concepts to my approach arises because the idea of 'a practice' is essential to ensure an understanding of science which does not neglect social holism.

Inherent Problems

There are however serious problems with this simple, straightforward view of practices. Recent work has shown that, if practices are taken to be kinds of entities shared by individuals, or transmitted from one generation to the next, then it is very difficult, if not impossible, to have a self-coherent conception of them.[7]

The arguments involved can be appreciated more if seen as a generalised version of an argument concerning the particular case of language. Davidson has argued that there is no justification for taking language to be some kind of entity shared by individuals and transmitted from one generation to the next. Our usual view on language makes us believe that each party in a conversation participates to some extent in a shared theory, and each knows that the other shares this theory as well, and each knows that the other knows and so on. The hidden assumption here is that there is some kind of object, called a a common theory, or a language, or a system of linguistic conventions, that exists independently of the individual

---

[7]   D. Davidson, 'A Nice Derangement of Epitaphs'; S. Turner, *The Social Theory of Practices: Tradition, Tacit Knowledge and Presuppositions.*

protagonists of the language. According to Davidson, such an assumption is wrong. All we need, when trying to understand the situation of two people engaged in conversation, is to look at the individuals themselves. We should not go beyond them towards some presumed object common to both. Looking at the individuals will reveal that each has a *prior* theory indicating how one is prepared in advance to interpret the utterance of the speaker. Apart from this, each has also a *passing* theory which constitutes the way each in fact interprets the utterance of the other. Passing theories of one interlocutor constitute attempts at getting to what lies in the mind of the other interlocutor. As speaker and interpreter talk, their prior theories become more alike, and so also their passing theories. Complete mutual understanding occurs only as an asymptote when passing theories coincide. Speakers have the ability to converge on passing theories from utterance to utterance. Their starting points, however far back we want to take them, will usually be very different because it is very often the case that the way each acquired their linguistic skills is different. The strategies of each to bring about convergence will therefore be different as well, but this does not matter as long as convergence, and hence mutual understanding, occurs. Davidson therefore concludes that, by giving such a plausible account of mutual understanding, he has shown that the previously held idea that language is a quasi-object is not only dispensable but also mistaken: 'There is no such thing as a language, not if a language is anything like what many philosophers and linguists have supposed. There is no such thing to be learned, mastered, or born with. We must give up the idea of a clearly defined structure which language-users acquire and then apply to cases.'[8]

This kind of argument can be applied to the idea of practices, especially if these, just like language, are taken to be collective objects in which different protagonists, perhaps at different periods in history, share to some extent. Such an application has been carried out by Stephen Turner. I will mention two of his most convincing arguments showing that the notion of practice, at least on one understanding of it, is internally inconsistent.

First of all, just as Davidson detached himself from the usual claim that there is some hidden quasi-object, in his case language, which is shared by individual protagonists, so also Turner questions the usual understanding that individual patterns of behaviour are explained by alluding to the causal powers of some kind of quasi-object we refer to as a practice. The problem here is the familiar one of underdetermination of theory by data. Evidence of sameness as regards appearances is *not* evidence of sameness as regards the causes of these

---

[8] D. Davidson, 'A Nice Derangement of Epitaphs', p. 446.

appearances. If two persons act in the same way, one cannot conclude that what made them act in that way was also the same. They may have been educated by different methods. Think of the mastery of a language as a first language and the mastery of the same language as a second language: the final effect is the same, but the underlying causal ancestry differs widely. Quine employs the image of bushes that are clipped carefully to have the same shape. The external shape does not imply that the internal branch structure is the same. Tutoring and disciplining are means by which people are induced to respond in ways that are the same. But methods of tutoring and disciplining may differ. The sameness of behaviour is the only feature we can be sure about.

This observation shows that anyone who wants to claim that practices are kinds of objects that are transmitted from one generation to another, and are manifested by the similar actions and behaviour patterns of protagonists in each generation, is making extremely unwarranted assumptions. Such a person will be making the same kind of mistake as the one who thinks that two people who speak the same language have learnt it in the same way.

The second of Turner's arguments I want to mention involves the identity conditions of a given practice. These constitute a considerable problem. Examples of practices come in many different forms. For some social theorists, the notion of practice can be employed to refer to a particular way of walking, a way which can be distinguished by the locality of where it is engaged in, such as the way American girls walk, or the way Parisian girls walk. For others, practices are much more complicated, socially realised, organised actions, like architecture, fishing, scientific practice, and games involving teams and supporters and so on.[9] The question therefore arises whether agreement can ever be reached as to what are, at least, the necessary conditions for something to be a practice.

An affirmative answer to such a question seems too difficult to conceive, given the enormous number of possible variations of social and individual human action. Moreover, even if we consider what appears to be the simplest possible kind of practices, namely the ones corresponding to individual characteristics like, for example, the ones corresponding to individual ways of walking, we still have to face formidable problems of the following kind. The 'American gait' is understandable only when we have an idea of what a normal gait is. Our usual assumption is captured by Cicero's maxim: 'custom is second nature'. We assume that there is a natural way of doing something, and then custom is added on. We are

---

[9] An example of the former, mentioned by Turner, is M. Mauss, 'Body Techniques'. An example of the latter is A. MacIntyre, *After Virtue*, p. 273.

motivated to explain why some girls walk one way while others walk another way because we assume that we know how girls normally walk and therefore can identify what they add on to this: for decoration, as it were. But such reasoning is fallacious: we have no reason to believe that what we take as normal is also taken as normal by observers from other cultures. African observers will probably carve out the style of American girls differently from French observers: if African girls walk in ways similar to American girls, they will not mention gait at all. The taxonomy of practices, even in the simplest cases, cannot therefore enjoy any acceptable level of objectivity.

If I am to employ the notion of practice, with the aim of safeguarding social holism in my account of science, I am obliged to seek an understanding of 'practice' which is robust enough to withstand such objections.

A Satisfactory Account of Practice

The first argument depends on the clear distinction between what is external and what is internal, between what Turner calls phenotype and genotype. The only thing we can be sure about is what we see, namely the similar behaviour patterns of a number of people. Turner insists that we are not entitled to draw any conclusions about the cause of this similar behaviour. In fact, he remarks that in most cases, similar behaviour patterns have different causal chains. In this argument however, the assumption that a practice is a cause of behaviour patterns is unwarranted. Turner's argument is devastating only for those who start with the supposition that practices are kinds of entities that *cause* behaviour patterns. But there are ways of understanding practices that have nothing to do with considering them causes of behaviour patterns. Two persons may have arrived by different routes at the mastery manifested by their behaviour patterns, just like the case of the two speakers of English: the speaker of English as a first language, and the speaker of English as a second language. But a multiplicity of different routes of acquisition does not necessarily mean different thing acquired.

What we can learn from Turner's argument is to avoid seeking a notion of practice that pretends to refer to a cause of the observable behaviour patterns.

The second objection referred to the fact that identity-conditions for practices do not seem to be binding to all observers. What is a practice for one observer may not be the same practice for another. The argument depended on the lack of sufficient criteria that hold for all occasions of practices. If the essence of a practice cannot be defined, so as to show clearly what is and what isn't a practice irrespective of the observer, then the notion is vacuous.

The problem with this objection lies in the essentialist assumptions it brings in. The framework set by these assumptions makes us think that the taxonomy of practices happens as follows. We say that fishing is a practice consisting in taking the rod, going to the sea-side, discussing with friends, hooking up the bait, and so on. Then we ask the question: on what grounds are we bringing precisely these descriptions together and not others? The temptation is to say that, since we do group things together in this way, then there must be something to it: an essence or an ontological reality being referred to by the notion of practice. So we are faced with sub-options here: we can either give in to this temptation and look for the identity-conditions of the entity we call practice — in this way we are reifying practice to some extent — or we can resist the temptation and stick to a strictly nominalist view of practices, according to which what determines a practice depends only on what observers decide.

But the choice is not limited to being either essentialist or nominalist. There is the option of employing the notion of family-resemblance referred to before. Hence, as has been said about discourses, when we ask: 'what is common to all practices?', the answer is not a list of necessary and sufficient conditions for something to be a practice, a list constituting the essence of a practice. Nor is the answer the fact that they are all *called* practices. When faced with a limited number of examples of practices, we can indeed identify common features. But when considering the entire set, the answer suggested by the family-resemblance model consists in saying that what is common to all is the fact that they *are* all practices. If we take the concept of practice to be a family-resemblance concept, then we are introducing borderline cases: how much resemblance to other cases is sufficient for something to be a practice? But this blurring of the edges does not prevent a concept from functioning successfully. The concept of 'game' is used successfully. We have no reason to say that other family-resemblance concepts like 'practice' do not. We have no clear reason either to assume that all entities referred to by our language are free of vagueness irrespective of our way of representing and talking about them.[10]

What Turner's argument underlines is the obligation to avoid — on pain of procuring considerable problems — giving an essentialist version of the concept of practice. We should avoid seeking a list of essential properties or a definition of practice consisting of necessary and sufficient conditions for something to be a practice. Given my option of working within the family-resemblance model, the best I can aspire to is a statement that serves as an indication how to identify

---

[10] See R. Bambrough, 'Universals and Family Resemblances'; T. Williamson, *Vagueness*.

practices in most cases. The case here is similar to the search for convenient expressions that may help a child learn what a chair is. We can use ostensive definition involving a number of token examples. We can also use the expression: 'most chairs have four legs'. Ostension and expressions of this kind will enable the learner to grasp the concept and learn it. Such a procedure was discussed in some detail in the previous chapter.

Reflection on Turner's two arguments therefore have had the beneficial effect of leading us away from hazardous ground in our search for a plausible way of talking about practices. As a first attempt at giving a useful statement about practices, I will suggest the following:

P   Most practices are collections of regular activities of one or many protagonists aiming thereby to achieve a purpose.

Further refinement of this tentative statement will now be carried out as regards three main issues: the first concerning habits, the second rules, and the third purposes.

People who do certain actions regularly are often described as having a habit. In general, habits may be understood as dispositions to act in a certain way, and are usually acquired by frequent repetition of the same action until it is almost involuntary. When considering a practice involving many protagonists, there is an observable similarity in external behaviour of all protagonists. This should not be taken to imply that all protagonists are participating in one collectively shared singular habit which enjoys some kind of existence independently of them. Such an implication would make practices into objects or quasi-objects, and the consequent idea of having such objects transmitted from one generation to the next has to face the insurmountable problems of the location of such objects, of their sameness over time, and so on. Suppose we start with something literally the same, for example the same treaty is signed by two parties. Once the treaty is read, the sameness ceases. Texts are subject to interpretation. If one says that the interpretation of texts is subject to tacit conventions that assure sameness in reception of the message encoded in the text, then one just moves a step further. One still has to show how these conventions, if they exist, are in fact the same for all hearers. The only way one can check that the text has been understood in the way one should understand it is to observe external behaviour. But education and discipline both act directly only on external behaviour. One has no reason to believe that the internal convention for interpretation is the same when the external behaviour is the same. To avoid these pitfalls, Turner takes the option of considering only habits as regards *individual* persons:

> By performing in certain ways, people acquire habits which lead them to continue to perform, more or less, in the same ways. The observances, so to speak, cause *individual* habits, not some sort of collectively shared single habit called a practice or a way of life, which one may possess, or fail to possess.[11]

Moreover, even about these individual habits, Turner's views are minimal. He refrains from claiming anything about them other than that they result in the same outward observable behaviour. This is because different protagonists may arrive at the same behaviour patterns by going through different ways of acquisition: some may have learnt primarily through written formulae, others through imitation of the actions done by experts, and so on.

This way of arguing however does not make the right distinction between what is acquired from the way one acquires it. As has already been pointed out, that two people acquire something in different ways does not necessarily imply that what they acquire is different. Hence I will distance myself from Turner's views by suggesting that the individual habits involved in a practice are of the same kind, namely of the kind that give rise to, or is manifested by, such and such behaviour. By claiming this, I am not committing myself to the existence of any direct causal link between habit and action. All that is necessary here is to say that a specific action is a *manifestation* of the particular disposition we call habit.

Some may object perhaps that my move here is hopelessly circular because no proper explanatory role is being played by habits. In fact, to say that the behaviour patterns of two people are the same because they have the same habit is apparently as useless an explanation as saying that a certain potion makes you sleep because it has a dormitive virtue.

This objection however can be responded to in the following way. A certain pattern of behaviour which is similar for different people is often the result of various causal chains having different starting points for different people. Hence methods of tutoring and disciplining in schools may be different even though the external behaviour of the students is the same for each student. Let us call the common behaviour B and the different causal chains that may lead to it C1, C2, C3, ... Turner takes the position according to which the notion of habits should be

---

[11] S. Turner, *The Social Theory of Practices: Tradition, Tacit Knowledge and Presuppositions*, p. 100.

limited to the individual. Hence, although the outward behaviour is the same, each student will have different habits according to the causal ancestry of that outward behaviour. For Turner, there is no distinction between the habit of an individual and the causal ancestry of the behaviour pattern of that individual. Hence for him habits are causally relevant, and therefore supply proper explanations.

In my view, a habit is not identical to the causal ancestry as regards an individual student. The disagreement between me and Turner depends on where we put our emphasis. The situation we are both faced with is one where the same behaviour patterns are produced by different causal chains. He emphasises the difference in causal ancestry and concludes that there is no reason to believe that there is anything similar in the people acting in that way, apart obviously from the fact that they act in that way. As opposed to this, I want to emphasise the similarity between the actual behaviour. I emphasise this by saying that there is no reason to believe that there is a different thing acquired. In fact, the similarity gives some plausibility to the claim that there is some property or disposition in common to all persons involved. This disposition is manifested by the similar behaviour. Hence, in my view, a habit is not identical to the causal ancestry as regards an individual student. Yet I want to claim that using habits in this way can supply proper explanations. To justify this claim, I will allude to what Frank Jackson and Philip Pettit call the program model of causal explanation.[12]

It is not implausible to hold that a higher-order property is relevant to a certain effect. Hence the elasticity of the eraser is relevant to its bending because elasticity is related to lower-order properties which are causally relevant. It is not vacuous to say that the bending of an eraser is explained by its elasticity. It is not vacuous because elasticity is related to the molecular structure of the eraser. The elasticity brings about the bending in so far as the molecular structure brings it about, and not in virtue of an autonomous power. But if the molecular structure has done the explanatory work, why does one need the elasticity? What is the specific contribution of elasticity in the explanation?

We may call the elasticity the dispositional state, and the molecular structure the realiser state. Notice that I am taking the dispositional state to be distinct from the realiser state. I am not saying that elasticity just is the molecular structure. If I take that option, then there certainly will not be any contribution of elasticity in the explanation. If the distinction is kept however, then the claim is that the bending of an eraser can indeed be explained by saying that it is elastic. This is an acceptable

---

[12] P. Pettit, *The Common Mind*, pp. 32-42; F. Jackson and P. Pettit, 'Program Explanation: a general perspective', pp. 107-117.

explanation because the dispositional state, the elasticity, 'programs' for the bending: it is saying that the eraser has a *certain kind* of molecular structure and not another. Consider another example: a glass container cracks when boiling water is poured into it. Why did the flask crack? One simple answer is: because of the boiling water. It was not because the supporting clip was too tight; it was not because the glass gets more brittle as years go by and this flask was quite old. It cracked because of the boiling water. A lower-order property can also supply an explanation: the flask cracked because of the momentum of this particular water molecule was enough for the breaking of the first bonds in the molecules of the glass surface of the container. Both these answers are relevant. Both give proper explanations. Neither can replace the other completely and neither is vacuous like the one involving the dormitive virtue. Pettit explains this in the following words: 'The boiling state of the water ensures a distribution of moving molecules such that there is almost bound to be some molecule that has a momentum and position sufficient to crack a molecular bond in the surface. The realising distribution does not produce the cracking but part of it does: this particular molecule. The boiling state is causally relevant so far as it more or less ensures the presence of such a productive factor and thereby programs for the cracking.'[13]

My suggestion is that this kind of model of explanatory relevance can show how using habits in explanations is not vacuous. Replace the dispositional states in the previous examples by the habit, and replace the lower-order state by one of the causal chains leading to the specific behaviour pattern we are considering. Hence, instead of elasticity and instead of the boiling water we now have a habit. Instead of the particular molecular structure of the eraser material and instead of the momentum of this particular water molecule hitting this particular surface molecular bond we now have one of the causal chains: either C1, or C2, or C3, ... So let us take one particular element of behaviour in need of explanation. Suppose we ask: why does this student always start the day by doing stretching exercises? One straightforward and simple explanation is: because he has a habit. It is not because he is worried about anything in particular; it is not because he thinks his neighbour is using a telescope to watch him to check that he is regular. He does it because he has a habit of doing it. A lower-order property can also supply an explanation: the student behaves this way because he has gone through a causal chain C1, consisting of, say, the way his mother used to make him wake up early and go to bed early, and the way this was reinforced by the kind of boarding school he used to attend as a teenager, and so on. In line with the program model discussed

---

[13] P. Pettit, *The Common Mind*, p. 39.

above, it is plausible to hold that both of these answers are relevant explanations. Neither can replace the other completely and neither is vacuous. When we say that he has a habit, we are saying that one of a particular set of causal chains is operative. We are saying effectively that either C1 or C2 or C3 or ... is in operation.

This shows that it is possible to use habits in proper explanations. In some situations, explanations using higher-order properties like habits are not only possible but advantageous. I have two main reasons for this claim, especially as regards the use of habits.

First, given the complexity of the causal ancestry of behaviour patterns, it seems plausible to hold that dispositional-state explanations play a more significant role in the habit case than in the boiling water case. When it is behaviour patterns that are to be explained, the intricate causal chain leading to the effect we see in the person's way of acting is difficult to determine, if it can be determined at all. The interaction that life in society involves makes it very difficult to conceive of a determinate causal chain leading to a behaviour pattern. When it is particular empirical events that are to be explained, sometimes our scientific models allow us to consider ideal situations where any intruding causes are minimised. We consider shielded environments where most variables are considered constant, or irrelevant. Hence, we can easily picture to ourselves a molecule with a certain momentum hitting a particular surface molecular bond. But it is more difficult to isolate a similar picture consisting of the causal ancestry of the element of behaviour consisting of doing stretching exercises on waking up. The conclusion to be drawn is that explanations using higher-order properties become more and more acceptable as the complexity of the *explanandum* becomes more and more difficult to handle.

The second reason that gives me confidence in claiming that explanation in terms of habits can sometimes be preferable is the following. Physics is often seen as a model of what proper explanation should be like. But even explanations in physics do not always give preference to explanation in terms of lower-order properties over explanation in terms of higher-order properties. It is true that most explanations in physics do involve an attempt at reaching an explanation in terms of a lower-order property. Such explanations seem more convincing, and therefore more worthwhile. However, we should recall the interesting and significant change in the status of explanation from classical to modern physics. On one interpretation of quantum mechanics, not all changes require explanation. Discontinuous action, annihilation of elementary particles, whether stable or unstable, and the radioactive decay of nuclei are all taken to be basic. They need no explanation. In classical mechanics higher-order relations could be reduced to lower-order relations. Many statistical relations were shown to be equivalent to the causally connected propositions of mechanics. Thus, in the kinetic theory of gases, pressure and temperature can be reduced to average values of the independent variables of the molecules. In this case, a higher-order property is simply a term which is used when we are ignorant of all the micro-properties that give rise to the effect we want to study. For the Copenhagen interpretation of quantum mechanics however, higher-order properties are not reducible in this way. Higher-order properties are not terms used when we are ignorant of hidden causal chains that would, if known, give the definite explanation. Higher-order properties are all there is to say: they supply a complete description. What is relevant to say about a radioactive substance is that, after a specific number of years, half of its atoms disintegrate. The question why this atom rather than that atom disintegrates first is considered irrelevant. This situation in this branch of modern physics may be understood as follows. Explanations in terms of lower-order properties are not needed anymore in certain domains. All the explanatory work is done by higher-order properties.

Even in classical physics, explanations do not always give preference to explanation in terms of lower-order properties over explanation in terms of higher-order properties. As an example, consider the scientific treatment of problems involving vibrating strings.[14] We study the resonance patterns by assuming that the string is a continuum and that its position is a continuous function of time. Given our other well accepted beliefs about the world around us, we should also say that this assumption is false, because strings are not continua. Strings are a very large number of molecules held together by binding forces of various kinds. Explaining vibrating strings by making the fictitious assumption that they are continua is explaining them by alluding to a higher-order property, namely the property that they look like continua as far as vibrating is concerned. This move is not only useful but indispensable. It is indispensable because, if we descend to the level of analysis of the string in terms of its lower-order properties, then we have to deal with a problem of billions of particles in mechanics. If one concedes that this problem in mechanics can be handled, which is obviously no small concession, it is practically certain that the resulting explanation will be relevant to the original questions about string-vibration only in so far as the structure of billions of molecules approximate a continuum. Moreover, analytical techniques can be possible in terms of higher-order properties and impossible in terms of the lower-order properties. We can speak of the derivative of the string function and examine the causal links this derivative has with other facts we know about the string. Such

---

[14] The example is from A. Garfinkel, *Forms of Explanation*, p. 166.

claims will be possible only if the higher-order properties rather than the lower-order properties are taken into account.

The moral here is that even explanations in physics, a discipline which is often seen as offering a model of what proper explanation should be like, do not always give preference to explanation in terms of lower-order properties over explanation in terms of higher-order properties.

To recapitulate, therefore: this first point of refinement of **P**, the notion of practice developed in the preceding paragraphs, consists in the following claim. The use of habits, even if understood as distinguishable from causal chains, is legitimate and perhaps sometimes preferable in explanations in general, and in an account of practices in particular. A habit is best seen as an acquired disposition over which we still have some control. It makes no sense saying that a token behaviour pattern is *directly caused* by the habit. In general, dispositions do not directly cause anything: the solubility of salt, say, does not *cause* the disappearance of salt crystals in water. It is *manifested by* the disappearance of salt crystals in water. This does not mean that dispositions are causally irrelevant. They are indeed causally relevant because they are a first indication of what is causally efficacious. We should hence say that the habit is *manifested* by a token behaviour pattern. We detect dispositions, just as we detect other things, by observing certain events. For a disposition like the fragility of a glass, what shows the fragility is the event of the glass breaking on dropping. For a habit, say Peter's habit of walking the dog every evening, what shows the habit is the event of his doing so on a regular basis, or the event of him becoming grumpy and restless if he stays indoors all day. The correct way of talking about habits is in terms of manifestations and activating conditions not in terms of their reducibility to causal chains.[15]

The second point of refinement concerns rules. The statement **P** about practices given above alludes to regular activities. This kind of vocabulary should not make us think that identifying a practice involves a process whereby one sees what kinds of actions are in accord with some given rule. This kind of mental matching-operation is very dubious. It seems important to emphasise that a rule is not a function in the brain, or a stencil allowing some actions to go through rather than others, as if the function or stencil can be defined without referring to these actions. A rule is inseparable from the actions that are in accord with it. The non-analysability of the relation between rules and actions in accord with them can be

---

[15] I am here endorsing a realist view of dispositions in general, and of habits in particular, as defended in D.H. Mellor, 'In Defence of Dispositions'.

expressed by saying that the relation is 'internal'. A relation between two objects is internal if it is unthinkable that these two objects do not stand in this relation.[16]

The argument showing that a rule cannot be formulated independently of the acts that are in accord with it depends on the simple observation that to discuss a rule at all, I must have an expression of it. And if I have an expression of a rule, I necessarily have a description of the acts that accord with the rule. When I 'see' the rule, I always 'see' the acts that accord with it. Some may object by claiming that when we say 'the rule of addition' we are not saying anything about what acts accord with it. But such an objection is futile: it is neglecting the fact that 'addition' is *just a short way* of saying something else. It is a short way of saying 'the act of putting together two or more numbers to find a number denoting their combined value'. This long statement is itself a short way of expressing the acts of adding in all their detail — it is a short way of expressing all the processes that a school teacher has to teach primary school children. Hence ultimately, a rule always employs the same symbol as the thought of the act which accords with it. From such considerations, it follows that it is inconceivable that I describe 'the rule that X' without using 'X'.[17]

This can be understood also if we refer to the previous account of rule-following. In the last chapter I mentioned how the Kripke argument shows that no finite set of examples can instantiate a determinate rule. Any finite set of examples instantiate an infinite number of possible rules. Hence just a finite number of additions done in primary school cannot be considered enough to determine the rule of addition. If a child who has learnt how to add looks to new cases, and asks whether they fit the rule or not, that is to say whether the answers to these sums are correct or not, then the positive or negative inclination that they induce in her will present itself as a disposition to judge yes or no for each case. As she considers the new cases to see how they fit with the rule, she has to see whether they present themselves in a positive or negative light, whether the actions she is used to together with the inclination produce these answers to these sums or not. The role of the inclination in the determination of a rule cannot be made redundant. It is the inclination plus the examples that exemplify the rule. The children in primary school are given examples of addition so as to foster the correct inclination. The

---

[16] I am using Wittgenstein's terminology: *Tractatus* 4.123.

[17] After going through the exegetical studies of G. P. Baker and P. M. S. Hacker, especially in *Wittgenstein, Rules, Grammar and Necessity*, I think that this is the argument that Wittgenstein presented in his rule-following considerations. It is a parallel argument to the one involving expectation (*Philosophical Grammar* § 134; *Philosophical Investigations* § 452 ff.).

rule is manifest to the agent, as an inclination and a set of exemplars. She will be able to target or address the rule: she will be able to think of it ostensively as *that* rule. But in so doing she has to refer always to some acts and the associated inclination.

The upshot is that one cannot formulate an approximate description of practices by assuming that there is a distinction between rules and acts. If we say that practices are rules which govern behaviour of some agents, we would be falling into this kind of error.

The third and final point of refinement concerns purposes. Practices usually concern a multiplicity of individuals. Even in the case discussed earlier of the American gait, which apparently involved only one person and how she walks, it is arguable that any notion of practice employed here covers how the great majority of American girls behave, rather than how one particular girl behaves. If this is accepted, the inclusion of the idea of purpose must be understood as referring to a collective purpose.

But can one have a reasonably coherent understanding of a group purpose? To answer this question, I will recall some important elements from the discussion concerning group belief carried out in section three of the last chapter. An extremely useful, initial distinction was the one between belief and acceptance. I worked there on the assumption that a belief that p is a disposition to feel it true that p. Acceptance that p is to treat it as a premiss that p. Hence belief has to do with feeling, while acceptance with reasoning. Here, a similar fundamental distinction must be kept in mind, this time between desire and purpose. On the one hand, I have a *desire* that p when I have a disposition to yearn or wish that p. On the other hand, I have a *purpose* that p when I adopt p as a goal. Hence, desires, like beliefs, have to do with emotions. Purposes, like acceptance, have to do with reasoning. Desires and beliefs arise in us slowly or suddenly, not as a direct result of our own will. Purposes involve the deliberate intention to achieve something. Obviously, any single person usually has many desires and purposes, and some of them may be conflicting. To have only desires but no purposes is to be completely impulsive, like an infant. To have only purposes but no desires is to be completely emotionless, like a guided missile.[18]

This distinction allows me to show that one can legitimately talk of group purposes but not of group desires. In section three of the last chapter, I showed that, if we insist that groups have beliefs, we have to face the heavy burden of explaining how a group can be said to have a mind of its own. To avoid this, I argued that the

---

[18] L. J. Cohen, *An Essay on Belief and Acceptance*, chapter 2.

best option is not to think literally of group beliefs but of group acceptance. Similarly here, if we insist that groups have desires, we have to face the same burden of explaining how a group can be said to have a mind of its own. To avoid this, one should not think of group desires but of group purposes. Hence when someone speaks of a group having an aim, or as being morally or legally responsible for its actions, he or she should not be taken to be referring to the group's desires but to its deliberate purpose. Similarly, in speaking of practices, one should not be taken to be referring to group desires but to deliberate purposes.

Moreover, we should not limit our analysis of group purpose only to well-defined situations where a committee decision deliberates and fixes an aim to be followed by a well-defined group. For our analysis to be useful in understanding science, we must allow for groups whose boundaries are not strict, and for groups containing some members who consider other members as experts. In such cases, the collective purpose does not have to be known explicitly in all its details by all the people engaged in achieving it. It often happens that some of those engaged in a particular activity are imitating the behaviour patterns of some experts prominent within that activity. This imitation is engaged in without knowing all the details of the purpose behind those patterns. In these cases, there is at least tacit agreement that the purpose of the prominent protagonists is being regarded as the collective purpose of all those engaged in that activity. In such cases, those who are not classified as experts can be said to know the collective purpose only as a vague statement.

Given these three points of refinement, I can now formulate a plausible understanding of practices that covers most cases I will be concerned with in my overall argument:

P*   In the majority of cases, a practice consists of actions grouped together (1) because they are geared towards the achievement of a collective purpose, and (2) because they are performed in a certain way which manifests habits.

The statement allows one to hold that practices enjoy some identity over time because, if all people stop acting for a certain period of time, their dispositions to act in certain ways can still be said to remain present — the presence of a disposition does not depend on the occurrence of any of its manifestations. Moreover, the statement underlines the fact that a description of a given practice may be good or bad, accurate or inaccurate, just like a description of any other entity. This is so because the collective purpose can be known vaguely or precisely, depending on the information available to the one describing the practice.

So, to sum up this section on the notion of practice, it is useful to recall the main steps in the line of argument. The main objection against the usefulness of this notion consisted of two problems. First, it was argued that evidence of sameness as regards appearances is *not* evidence of sameness as regards the causes of these appearances. Second, it was shown that the taxonomy of practices, even in the simplest cases, cannot apparently enjoy any acceptable level of objectivity. These lines of criticism were shown however to be less effective in undermining the use of the notion of practices than one would at first imagine. Against the first problem, I argued that a multiplicity of different routes of acquisition does not necessarily mean different thing acquired. So a lot depends on where the emphasis is put: either on the difference in the possible causal ancestry or on the sameness of the behaviour patterns. And both kinds of emphasis are plausible. Against the second problem, I argued that the real source of trouble lies in the underlying essentialism whereby one is tempted to give up the use of a concept if no set of necessary and sufficient conditions can be found for its identification. Adopting essentialism is however not an obligation: there are other options.

Given these reflections, one can go to the extreme of holding that the use of the notion of practices should be avoided at all costs, and moreover that habits should be defined as the *particular* causal chains of *particular* people. But since the convincing power of the problems has been considerably weakened, one can argue that going to this extreme is not justified. As regards habits, I argued that a habit is better understood as a disposition of a person, rather than as a causal chain. Therefore, we can still meaningfully understand a habit so as to allow that, if two persons have the same outward action, they may have different causal chains leading to that action, but they have the same habit. The bonus here is that, when we say that a person has a habit, we are saying that one of a particular set of causal chains is operative. As regards practices, I concluded that one is entitled to work with the assumption that practices often consist of actions with a certain purpose, actions manifesting habits.

This is the gist of the second section on the notion of practice. The previous section was on discourse. It concluded with the claim that, even though discourses should not be considered to form a family in virtue of features common to all, one particular feature is of particular importance for my overall aim, namely the possibility of agreement and disagreement. It is very probable that discourses having this feature are involved in scientific activity. The notion of Cognitive Command was brought in to elaborate on this attribute of discourses.

The investigation in the two sections thus achieves the aim of the chapter which was to analyse two important holistic concepts that facilitate the inclusion of holism in our understanding of science. One which is more related to cognitive holism is the notion of discourse. One which is more related to social holism, is the notion of practice. The next step is to take what was said in this chapter in a general way and apply it to the special case of science.

Chapter Five

The Case of Science

My aim in this chapter is to offer some arguments in favour of the claim that science can be viewed, in some respects, as a single whole. The consequence of this is that it is neither erroneous nor vacuous to speak of scientific practice, or of scientific discourse, in the singular. I will therefore be opposing a trend that can be discerned in some recent studies.[1] These emphasise the immense diversity of methodologies in operation in the various branches of what we call science. Such a diversity tempts us to acknowledge that one can speak of scientific practices in the plural, to refer, for example, to specific instrumental skills, but one cannot speak intelligently of 'scientific practice' in the singular: neither as the common factor of all these individual practices nor as the combination of all of them. Moreover the very idea of identifying methodological precepts or other kinds of descriptions that cover the entire range of all the sciences seems to imply that a simplified algorithm can capture a vast number of manifestations. Historical and sociological studies have given ample evidence that all such algorithms fall short of accounting for the extraordinary diversity of procedures pursued in science, even though all are claimed to be, in one way or another, expressions of Francis Bacon's inductive method.

I will proceed by applying the results of the previous chapter. The first section will concern the application of what has been said about practice to the case of science, so as to identify the sense in which one can legitimately speak of scientific practice in the singular. The second will concern what has been said about discourse and examine its application to the case of agreement and disagreement in science. This will lead to the third and final section which examines in more detail the role of experimentation in the picture of scientific practice and discourse described in the previous two sections.

---

[1] For example N. Jardine, *The Scenes of Inquiry*; S. Woolgar, *Science: the very Idea*.

1. Practices, Interests and Purposes

It was concluded in chapter four that a practice is generally described in terms of a purpose. If actions are engaged in to achieve a purpose, one can identify a practice. For the case of groups consisting of scientists, collective purposes of this kind have been called interests, especially by those engaged in sociological studies. In general, the claim here is that social, political, professional and other interests play a primary role in scientific innovations. There seem to be two main problems with such an approach. If such interests are described in very general terms, the role that they play in any explanation will risk becoming completely vacuous. The second problem is that, to avoid over-simplification, one cannot work with a view according to which scientists at one period of time have certain interests, and then, at another period of time, they deliberate which actions are in line with these interests. What we should expect is that interests and actions geared to achieve these interests are mutually interactive in the sense that interests determine actions but also *vice versa*. To solve these problems, detailed studies of particular periods of history have been carried out with the result that interests are described more modestly: they are defined in a way that makes them both specific and local.[2] The examples given by Nicholas Jardine of such interests are the following:

> interests in the demarcation and integrity of experimental natural philosophy, of its competent practitioners and its proper places of public performance; interests in the control of interpretation and explanation of experimental findings; interests in the preservation or renegotiation of the boundaries and statuses of existing disciplines and forms of knowledge; interests in the institution and enforcement of standards of moral decency in dispute and criticism; interests in the recruitment of potential allies and in the isolation of implacable opponents; interests in the preservation of structures of moral, religious, social and political authority against sectarianism and popular enthusiasm.[3]

The goal pursued in this section, it will be recalled, is to give a plausible account of scientific practice. Applying the results of the previous chapter, one may hold that scientific practice is the set of all actions and behaviour patterns geared towards the achievement of a certain scientific purpose. If such a purpose can be identified, then the legitimacy of scientific practice is secured. But the above list of

---

[2] For example, S. Shapin and S. Schaffer, *Leviathan and the Air Pump: Hobbes, Boyle and the Experimental Life*.

[3] N. Jardine, *The Scenes of Inquiry*, p. 180.

various interests may look, at first sight, rather worrying. It seems to indicate that the search for an overarching purpose of all those engaged in science is bound to fail. Scientific purpose is hopelessly fragmented. But a closer look at the issue will show that such fears are not justified. I will concentrate on two problems. The first one will concern how to understand the fact that protagonists of scientific activity seem to have different purposes. The second, which will demand a much longer treatment, will concern how to understand the fact that protagonists of scientific activity seem to have a purpose that is simply too general to be relevant.

The Problem of Multiple Purposes

That protagonists of a given practice may have multiple purposes cannot be denied. One may distinguish two main views on how one particular practice can be associated with multiple purposes. According to the first view, a group of people are considered as having an initial, well defined single collective purpose, for the achievement of which they engage in certain actions and show certain behaviour patterns. They thus engage in a practice. The members of the group then realise that to achieve their purpose they need to achieve other subsidiary purposes. Actions and behaviour patterns associated with each of the subsidiary purposes will constitute, according to the discussion carried out in the previous chapter, practices in their own right, and may therefore be called subsidiary practices. This view may be said to give an adequate account of the ever increasing ramification of science into different branches. Hence we have scientific practice dividing into subsidiary practices like the practice of biochemistry, the practice of solid state physics, and so on. In this view, however, there is an element of naïvety: it makes absolutely no allusion to the mutual interaction between actions and purposes mentioned above.

The second view includes this. Here, members of a group are taken as people having an idea of a collective purpose which is initially not definite but vague. They nevertheless engage in actions to achieve this purpose. Their engaging in this practice results in the discovery that the general purpose they thought they had needs refinement. It includes other subsidiary purposes. The full import of the original purpose is not assumed to be available to the practitioners at the beginning. This view gives a better account than the previous one of what happens in science. The interests enumerated in the above quotation are all examples of subsidiary purposes that are often not included in the formal methodological aspirations expressed in standards of scientific rationality. They constitute, however, purposes intertwined so closely with the general purpose described by such official standards that it is hardly conceivable how science as we know it can survive without them.

The awareness of the presence of such interests is arrived at by the group through the very process of engaging in the practice itself.

But how is one to conceive of the same set of actions that are achieving two different purposes? Do we have two practices or one? To answer this question, one has to recall that the way I have been describing purposes and interests allows me to hold that wanting to do A and wanting to do B is equivalent to wanting to do A and B. Hence no problem should arise if one practice is identified in terms of one purpose and then this purpose is discovered to be multi-dimensional.

The Problem of Too General a Purpose

Doubting whether there is any purpose at all associated with whatever we refer to when we say 'science' is unreasonable. Few would deny that being rational, and being scientific, and having an aim are intimately related. It is well known that some philosophers have made the formulation of the purpose behind scientific practice their main object of investigation. Karl Popper's main contributions in this area can be seen as consequences of his credo: 'when we speak of science we do feel, more or less clearly that there is something characteristic of scientific activity; and since scientific activity looks pretty much like a rational activity, and since a rational activity must have some aim, the attempt to describe the aim of science may not be entirely futile.'[4]

Can we suggest a rough idea of what constitutes the general purpose that gives a reasonably good description of scientific practice? If the purpose is described in very general terms, like 'to survive' or Aristotle's 'natural desire to know', or 'to arrive at the truth', then it will be difficult to distinguish between scientific practice and many other kinds of practices having to do with rudimentary actions people do habitually in everyday life. Science is however so complex that such general versions of purpose seem inevitable. From here one can already notice that a boundary between scientific practice and what may be called common practice, defined in terms of the most basic purpose of survival in ordinary circumstances, may not be as strict as is sometimes assumed. This point will come up later. But something can intelligently be said about the specific purpose of scientific activity, even though the notions used to describe such a purpose may not have strict boundaries. The way I will proceed to formulate such a purpose consists in three steps: first I will allude to rules of scientific method, then I will highlight the need for a historical approach, and finally I will show how the rules suggested by

---

[4] K. Popper, *Objective Knowledge*, p. 191.

methodologists of science are still valuable if understood in a certain way. After taking these three steps, I will be in a position to give an account of the purpose of scientific activity which is convincing and not too general.

The first step, therefore, is based on the observation that, to say something specifically about the purpose associated with science, the most plausible domain to turn to is the domain of methodology. It is here that the character of science as opposed to other forms of inquiry has been examined in detail. Methodology is the study of method. In general terms, it concerns the rules and evaluations that determine the interaction between evidence, argument, hypothesis and explanation. Sometimes methodology is used to refer to the list of rules. But the task of methodology includes also making sense of the rules it advocates and showing how they contribute to the rationality of science.

As a good paradigm example, one may consider Bacon's principles of inductive method. He suggested a method for the investigation of nature with the aim of gaining control and subsequent benefits. The practical success of a theory is the hallmark of its truth. His procedure involved the identification of the sources of error: it is false that the human senses are the measure of all things; we all have our predispositions and prejudices that are liable to mislead us; our language is also a possible source of error; the received view from previous philosophers should, if not rigorously justified, be considered false. According to him, to avoid these errors, one should conceive of causal hypotheses, which he called the forms, and then bring these to the test of experience. To conceive of the right kind of forms, one has to follow a rigorous and systematic collection of data. In line with most commentators, one may say that Bacon was committed to the simple claim that if theories are successful then we can be certain of their truth. The methodology he tries to defend therefore seems to be a version of the hypothetico-deductive method. The main purpose of the scientific enterprise is to search for well confirmed hypotheses. The hidden assumption is that knowledge rests on positive though inconclusive evidence. Scientists need to justify their hypotheses but they must also do this as fallibilists.

It is useful to view this in conjunction with another account of the methodology involved in science, a methodology which is also fallibilist but non-justificationist. In such cases, evidence operates only negatively. Falsificationism is this kind of methodology. It is based on conjectures and refutations as elucidated by Popper. Scientific activity should be directed towards the formulation of conjectures which are subjected to the severest imaginable empirical tests, in the hope that if they are false there falsity will be revealed. This procedure is more effective the more falsifiable are the proposed conjectures. Unlike the methodology suggested by

Bacon, Popper's version uses evidence only critically to refute conjectures. The purpose of scientists should be to try to falsify theories and to consider those that survive testing as corroborated or closer to the truth than the ones which have been falsified. Fallibilism is always present. Scientists have to make mistakes before being able to correct them. The main attraction of this view seems to be the way it gives an account of the purpose of science in terms of perceptual faculties and theories in a naturalistic setting. The purpose of science is to anticipate the world by making attempts which will be weeded out if the world does not match them well enough or if the world itself changes.

So Bacon and Popper offer useful reflections on the methodology of science. Both accounts have been debated and some other versions of the methodology for science have also been suggested. But there are some features common to all accounts. It seems safe to say that nearly all agree that contrary evidence cannot be disregarded, and that *ad hoc* hypotheses must be used sparingly if at all.[5] The existence of such features which are common to many accounts, even when these accounts are as different from one another as Bacon's and Popper's, indicate that it may not be impossible to formulate a general purpose for scientific activity that is reasonably convincing, is common to all methodologies, and is not so general as to be void of substantial content. As a good example of this, one can start by considering Popper's general statement about the aim of science: 'I suggest that it is the aim of science to find *satisfactory explanations*, of whatever strikes us as

---

[5] The Lakatosian methodology may be considered as diminishing the importance of evidence contrary to a research program. But it does not claim that any such contrary evidence should systematically be disregarded. This methodology is in opposition to Popper's on the *weight* that should be put on contrary evidence: according to Lakatos, one counterexample is not enough to bring down an entire theory. Research programmes have two major parts: the hard core and the protective belt consisting of auxiliary hypotheses. When the predicted novel facts do not receive confirmation, a research programme starts to degenerate. Prolonged lack of confirmation shows that the protective belt is losing its function. However it is not sufficient to eliminate a programme solely on the basis that it presently appears not to be making empirical progress. The programme could be in a degenerating phase from which it will recover. Hence the competition between rival research programmes can continue for centuries, because there is no such thing as a crucial experiment for a final decision. A scientific revolution occurs when a progressive research programme supersedes its degenerating rivals. So the lag between emergence of a new research programme and its becoming empirically progressive shows that there is no 'instant rationality'. But this is not problematic. For Lakatos, definite criteria are always available. In fact, at any given moment, it could be rational to stick to one's own research programme even though it is degenerating, but it is certainly not rational to deny its poor public record.

being in need of explanation.'[6] This positive claim is linked up, in his view, with a negative claim concerning the aim of removing bad explanations from our way of understanding the world. For example, explanations of the kind 'the tablets put you to sleep because they have a dormitive virtue' should be weeded out: 'This explanation is found unsatisfactory because (just as in the case of the fully circular explanation) the only evidence for the *explicans* is the *explicandum* itself. The feeling that this kind of almost circular or *ad hoc* explanation is highly unsatisfactory, and the corresponding requirement that explanations of this kind should be avoided are, I believe, among the main motive forces of the development of science: dissatisfaction is among the first fruits of the critical or rational approach.'[7] Popper's views here offer a good preliminary account of the general purpose of scientific activity.

Nevertheless, not to accept this view uncritically, one may ask whether the universal aim of science could be expressed in terms other than those dealing with explanation. One can examine the various available historical and sociological studies of science so as to determine the other major possibilities as regards the overall purpose of science.[8] These possibilities can be placed under the following four headings: firstly, the purpose of science could be merely the description of the world, rather than the explanation of it; secondly, the purpose of science could be the attainment of power over other people; thirdly, the purpose of science could be the gaining of control over the forces of Nature, in the sense of exploiting it for our benefit; finally, the purpose of science could be expressed in simple hedonistic vocabulary, namely the enhancement of personal pleasure and the reduction of pain. This classification under four headings does not mean that it is logically inconceivable to do science for some other purpose. It means only that, as regards the normal understanding of science, most of what is said can be fitted into one or other of the above headings.

Now, as regards the first kind of purpose mentioned, one can readily understand that merely describing the world, although necessary for science, is not sufficient. The describing is done in view of something else. This something else is usually related to the prediction of some aspects of the thing or event in question. Thus one starts by describing a pulsar so as to see how it's nature is related to what we know about stars; one starts by describing a disease to see how to cure or prevent it. Science is already being done during the describing phase. But stopping short of

---

[6] K. Popper, *Objective Knowledge*, p. 191.

[7] *Ibid.*, p. 192.

[8] A useful guide here is L. Stevenson, H. Byerly, *The Many Faces of Science*.

offering an explanation on this basis is stopping short of doing science. The upshot for our argument here is that Popper's suggestion of taking explanation as the purpose of science is more plausible than taking mere description.

As regards the other possibilities, namely that the purpose of science might be the attainment of power over other people, or the gaining of control over Nature, or the arrival at a state of more pleasure and less pain, closer scrutiny will show that all of these involve explanation as a condition. Put simply, if we do science with the purpose of gaining power over other people, or control over Nature, or more pleasure and less pain, we must want also to explain. The implication does not, however, go the other way round: if we do science to explain, it doesn't necessarily mean that we want power over other people, or that we want control over Nature, or that we want more pleasure and less pain. This asymmetry shows that it is more accurate to take the purpose of science to be explanation rather than any ulterior and thus more general motive. Such more general motives, like the ones mentioned above, contain the need for explanation, just as a set of Russian dolls all contain the smallest. They are answers to the question 'What is the purpose of engaging in science?' rather than to the question which interests us here, namely: 'What is the purpose of doing these specific actions, which, when taken together, constitute scientific practice?' These three kinds of purpose are therefore not to taken as the purpose of science because they are too broad. The other possibility, namely describing the world, was also unsatisfactory, because it is too narrow. These conclusions indicate that the correct way of talking of the purpose of science is to follow Popper and say that it is explanation.

I come to the second step of my investigation. This concerns the need for a historical approach. What Popper should have added is more detailed treatment of the criteria which determine which aspects or events 'strike us as being in need of explanation' and which do not. This has to do with what kind of questions are considered relevant, or real, for a scientist. One way of approaching this topic has been suggested by Jardine and consists in holding that a range of questions are real in a given scientific community at a given time when they are questions which the members of the community can see how they can get to grips with.[9] Understanding questions real in my community implies that I appreciate what the community considers relevant to those questions. Understanding questions unreal in my community but real in another community involves appreciation of the considerations that would be taken in that other community to be relevant to that question. Just as we can have live metaphors and dead metaphors in literature, so

---

[9] N. Jardine, *The Scenes of Inquiry*, chapter 3.

also we can envisage real questions and unreal questions for scientific inquiry. For any given community of inquirers we may have questions which used to be real but lost their relevance through the years. Such questions, having been replaced by others, are now considered dead: they are usually taken to be totally misplaced, or the embarrassing sign of our previous ignorance. So a useful refinement of Popper's views on the purpose of science is the observation that what strikes us as being in need of explanation is governed by what questions are considered real by the community we are in.

To arrive at a formulation of a purpose which unites all scientific activity, the suggestion I am following is to take Popper's view consisting of the two claims (1) to find explanations of whatever strikes us as being in need of explanation, and (2) to be critical and weed out unsatisfactory explanations. I add however the important injunction that what strikes scientists as being in need of explanation is dependent on what questions are considered real. And this addition apparently opens up a kind of Pandora's box: the hope that Popper seemed to have of grasping what the purpose of science should be for all ages begins to seem unrealisable.

It is useful to dwell on this point a little longer. By analysing what motivating forces are acting on a community of inquirers when a shift of questioning occurs, one will see that Jardine's distinction between real and unreal questions is very useful but may need some refinement if it is meant to cover most cases.

Let us take a recent example. In the US, in the first week of October, 1993, the Superconducting Super Collider was effectively killed. It had been planned to be a proton collider that could open up a qualitatively new domain of physics and ensure the US of its lead in high-energy physics with respect to the European centres such as the large electron-positron collider (LEP) at CERN in Geneva. What was it that blocked this huge enterprise? Problems of poor management were certainly present, but the main reason apparently was the ever-increasing costs which were drawing from resources desperately needed for the economy.[10] Interesting insights into what arguments were used to prove that the SSC is essential to tackle questions which were then considered real can be obtained from a paper presented by C. Quigg at a Workshop in 1986. It seems that the three main *real* questions according to Quigg concerned the following points: (1) the SSC was conceived to take the step needed for a 'thorough exploration of the 1 TeV scale'; (2) the SSC 'will clarify the structure and symmetry of the fundamental interaction and allow us to extrapolate with greater confidence back to early times', in other words, it will simulate the conditions that prevailed about 10^{-15} second after the Big Bang; (3) 'with the

---

[10] D. Ritson, 'Demise of the Texas Supercollider'.

support of our government, hard work, and a little bit of luck, we may have, by 1995, a new instrument to explore the 1 TeV scale, and to bring us closer to the dream of an enduring understanding of all natural phenomena'.[11]

Jardine draws a distinction between real and unreal questions. This example shows that a question can somehow be 'in between' reality and unreality. Take the questions which are behind Quigg's statements. These were real for him and his group, because they were questions which the scientific community could, according the opinion at the time, come to grips with. What are we to say about these questions now, given that the SSC project has been abandoned? Is it a real question now to ask: what happens in the 1 TeV scale? The plausible way to talk of these questions seems to be, on the one hand, to hold that the scientific community has realised that these questions cannot be tackled given today's scientific, technical, economic and political possibilities; and, on the other hand, to hold also that these questions still strike us as needing explanation.

Admittedly, people following the line traced by Jardine will make the additional claim that the importance of a question which is considered to be impossible to tackle, given the possibilities at a time, will steadily diminish until the question loses its reality. It loses its 'striking' power. This loss may take time, given the capacity of inquirers to put questions 'on hold', as it were, in the hope that scientific, technical, economic and political possibilities will someday become favourable. But one must concede that historical studies have given many reasons to believe that such loss of 'striking' power does occur.

I am now in a position to bring out the consequence of these reflections in the context of my original aim. This aim was to fix an account of the purpose of science so as to justify that there is such a thing as one scientific practice distinguishable from other kinds of practices. The historical dependence which creeps in because of criteria of relevance demands that some opening is left to allow shifts in relevance from one period in history to another.

But this does not mean that a reasonably accurate and unified account of the purpose of science cannot be formulated *given a set of real questions for one period of history*. The easiest period of history to deal with is obviously our own, where it is most probable that we know about the questions that are real and those that are not. Other periods of history will be more difficult to handle. But the purpose of scientific activity can be described independently of this, namely by the three points: first, to explain what strikes us as in need of explanation; second, to eliminate unacceptable explanations; third, to do this according to some rules

---

[11] C. Quigg, 'The SSC: Scientific Motivation and Technical Progress'.

described in methodological studies. The point I want to make is that the account of the purpose of science suggested here is plausible even though we are obliged to leave ourselves open to shifts in reality of questions.

But what about the rules described in methodology? Isn't there a historical shift there as well? This leads me to the third step in my investigation of the purpose unifying scientific activity. If I am committed to defend the unity of purpose, shouldn't I be worried that the rules suggested by Bacon differ from those suggested by Popper, and that these in turn differ from those suggested by Lakatos, and so on? At this point, the temptation is to enter into a detailed evaluation of these different methodological rules so as to decide which set should be accepted. But, if one succumbs to this temptation, then one will, as far as our search for a scientific purpose is concerned, miss the wood for the trees. Although, in its own way, a detailed study of method is very important, one should not think that this kind of investigation is all there is to do. There are also meta-methodological considerations. Such considerations enable one to respond to the following crucial question: as philosophers of science shift from the acceptance of Baconian rules to the acceptance of Popperian rules, and then to the acceptance of Lakatosian rules, and so on, are they getting any closer to the presumably fixed, mind-independent rules of science? Such a question retains its importance whatever particular set of rules are accepted. It deals with the underlying assumptions of methodological studies, rather than with their content. There seem to be three major ways of answering it. First, we may accept that there is indeed a fixed set of rules defining the scientific method and declare, moreover, that we have in fact come to identify these rules: we have succeeded where our predecessors have repeatedly failed. As a second possible answer, we may hold that there is indeed a set of rules defining the scientific method but admit further that what we have to offer now is a fallible approximation. On this view, one set of rules, presumably the set we accept now, should be considered as closer to the true set than other sets suggested previously. As a third possible answer, we may reply to the question by denying the existence of any mind-independent set of methodological rules defining science. On this view, the contents of methodological studies are descriptions of well-established conventions. The first response is naive and considerably problematic, mainly because it assumes that our present views on methodology are in need of neither development nor correction. The other two however are both reasonably defensible. They both accept that our views on scientific method change in the course of history.

If such change is accepted, the status of the rules we acknowledge at any given time has to be described with care. The crucial question is the following: are accounts of methodology saying what *should* be done or what *is in fact* done? From the discussion of the previous paragraphs, it seems clear now that the value of these methodological studies lies primarily in the fact that they are *descriptions* of the way scientists conduct their scientific activities rather than rigid, a priori *prescriptions* of how they should conduct them in all the foreseeable and unforeseeable future. That such descriptions are done in retrospect should not be a worry to us: in the view taken here, protagonists may engage in a practice even if the collective purpose they have — the purpose which may be used by observers to unify the actions into one group — is not fully understood or fully explicitly statable by these protagonists. Hence my view is sympathetic to such studies of rules or scientific method, but does not give them the full a-temporal jurisdiction some would like them to have.

By denying this a-temporal jurisdiction, I am taking a position opposed to those who aspire to give a definite final version of the over-arching method defining scientific inquiry. Imre Lakatos for example could not accept that scientists possess what 'cannot be articulated and made available to the layman outsider'.[12] He seemed to commit himself totally to a statute-law approach as regards understanding changes in scientific world views: he wanted to legislate in anticipation of concrete cases of such changes by fixing the method and purpose of science once and for all. The position I am advocating is different. The purpose of scientists, consisting of a multiplicity of interests, is always in interaction with the activities engaged in by scientists and hence cannot be definitely and explicitly formulated in full. The boundary around whatever we refer to by the terms 'scientific practice' is not a clear dividing line. I want to give an account, therefore, without committing myself for or against the Lakatosian assumption. I want to defend a position which does not depend on the assumption that there is, or that there isn't, a general algorithm, waiting to be discovered, that generates once and for all the innumerable interests and subsidiary purposes that should be allowed in scientific activities.

My position corresponds to a case-law approach.[13] Case-law is the general term for principles and rules laid down in judicial decisions, and for generalisations based on past decisions of courts and tribunals in *particular* cases. What is fundamental in case-law is not that previous decisions are reported, nor that judges

---

[12] I. Lakatos, *The Methodology of Scientific Research Programmes*, p. 176. On this point see also: I. Lakatos, 'The Problem of appraising Scientific Theories: three approaches'; 'Popper on Demarcation and Induction'.

[13] The application of statute-law and case-law approaches for scientific activity is mentioned in A. F. Sanders, *Michael Polanyi's Post-critical Epistemology*, pp. 138-145.

and other adjudicators in later cases look to previous decisions for help or guidance, but that the previous decisions are treated as normative and looked to for rules which should be applied. Furthermore, the higher courts in reaching decisions do so in the knowledge that their decisions are laying down *strong indications* which will be followed in the future by later courts, as opposed to laying down *a definite system* of laws for the entire future. Similarly for the explicit formulation of the methodology of science: attempts at formulating methodological principles and rules are useful in so far as they are treated as normative to a certain extent for future questions regarding the activity of scientists, but one must always keep in mind that new activities engaged in by scientists may result in the uncovering, or in the highlighting, of other subsidiary purposes and interests that were not included in the previous formulation.

To put this more schematically, the case-law approach advocated here entails the following. We start with a situation where a number of cases can be studied. The cases include simple or complex crises in scientific practice: the problems that gave rise to them together with their resolution, each in a particular historical and sociological setting. A study of such cases may result in a consensus as regards the kind of rules that were followed in all these cases, even though these rules may not form a system. Different scholars may suggest different lessons to draw from such cases: hence the difference between a Baconian and a Popperian. Suppose now that we are facing a new case. Taking the case-law approach means that scientists facing the new crisis are not fully constrained by the set of methodological rules derived from the previous cases. The rules are not useless. They give indications. But the novelty of the case implies that the resolution cannot be fully derived from these rules. One plausible way how to understand the subsequent procedure is to hold that the rules established previous to the new crisis enable scientists to acquire a skill that enables them to face this new crisis and figure out a way of resolving it. In so doing, they will be giving indications for future generations how to handle further new crises if they turn up. They will be giving indications but not a definite algorithm.

A few remarks are in order on how best to understand skill. Take the classic example: cycling is a simple skill which concerns the basic problem of how to keep one's balance. The rule governing this skill may be stated explicitly: in order to compensate for a given imbalance, one must take a curve on the side of the imbalance whose radius should be proportional to the square of the velocity divided by the tangent of the angle of imbalance. This is the case obviously when air-resistance, friction, and other similar forces are considered negligible. To design a robot that is meant to ride a bike, knowledge of such a rule is indispensable for the designer. But when it comes to riding a bike ourselves, that is ourselves as distinct from robots, learning the rule does not turn us into expert cyclists. Most cyclists know nothing about it. Yet they are excellent not only when air-resistance and friction are negligible but also when cycling on rusty bikes in strong wind.

Having a skill therefore seems to be a case of one knowing more than one can tell. It is certainly untrue that cyclists know all the rules involved and it is extremely unlikely that they had sometime in the past acquired them during the teaching period and can retrieve them from memory, in explicitly propositional form, if they engage in prolonged introspection. Such inaccessibility of knowledge in a propositional form is often taken to show that we are dealing in these cases with a special kind of knowledge that may be called tacit.[14] This kind of knowledge may be said to be what distinguishes persons who know how to cycle from exact counterparts who do not. It may presumably involve the possibility of degrees of tacitness, something which may be observed in the learning of a skill, say that of playing a musical instrument. Some indications in propositional form are indispensable for the beginner, but later on the player has to shift his attention to what lies beyond these indications and become more and more independent of them, even perhaps to the point of reaching such a high standard of skill that he may go against the preliminary indications given by his teachers. The inevitable need for us to allow a role in our explanations for such tacit knowledge was forcefully illustrated by Gilbert Ryle.[15] He observes that, when considering two chess players, one clever and the other dumb, we can never conceive of a list of facts or rules that the clever player may impart to his weak opponent and that will necessarily make that weak opponent better. The dumb player may still know all the rules by heart and still be unable to apply them at the right time. Knowledge of how to follow a rule is not a case of knowing-that but a case of knowing-how: knowing a rule is realised in acts that are in accord with the rule, not in theoretical citations of it.

These reflections can be applied to the previous discussion regarding scientific activity. What one is led to conclude is that a scientist cannot be viewed as someone who is primarily an expert in knowing-that, an expert who knows a lot of facts. It is much more plausible to view a scientist as someone who is primarily an expert in knowing-how, as someone who knows how to go about asking the right type of questions concerning discovery in a specific context. On this view, scientists are therefore those who have acquired a particular skill. They are good at

---

[14] M. Polanyi, *Personal Knowledge*.

[15] G. Ryle, 'Knowing how and Knowing that'.

matching the demand of a particular task to their capacities. Ratiocination is not the only way of being rational. One way this can be understood is in terms of Charles Sanders Peirce's idea of the instinct for successful abduction. Abduction concerns the question of how we decide which hypotheses are worth testing. Peirce was led to the conclusion that evolutionary adaptation has given human beings an instinct for guessing right which enables them to make successful abductions. He argues that we had better abandon the search for truth 'unless we can trust in the human mind's having such a power of guessing right that before many hypotheses have been tried, intelligent guessing may be expected to lead us to the one which will support all tests, leaving the vast majority of possible hypotheses unexamined'.[16]

So, to summarise the line of argument followed up to now, it is useful to recall the original problem: if the purpose of scientific activity is defined in very general terms, then there will not be enough distinction between scientific practice and many other kinds of practices. The way I proceeded involved three steps: first, I examined briefly how the purpose of scientific activity can be extracted from the rules of scientific method. Then I emphasised the need for a historical approach: I included the fact that what strikes us as being in need of explanation may shift from one period of history to another. The third step involved a defence of rules of scientific method. That the historical approach should be endorsed in articulating a purpose for scientific activity does not mean that such rules are useless. They are still valuable if understood in a way characterised by the case-law approach. This approach brings to light the element of skill that protagonists of scientific activity possess. After taking these three steps, an account of the purpose of scientific activity which is not too general can be formulated as follows: (1) to find explanations of whatever strikes us as being in need of explanation, given a set of real questions; (2) to be critical and weed out unsatisfactory explanations; and (3) to do this according to a set of methodological rules whose role is not to fix the activity of scientists once and for all but to give indications and foster skill.

So this resolves what I called the problem of too general a purpose. Given that the previous problem, namely the problem of multiple purposes, had also been resolved, one can be confident that a unifying purpose of all those engaged in scientific activity can be identified. One is entitled to say therefore that there is such a thing as one scientific practice. Referring to the concluding general statement about practices at the end of the last chapter, statement **P***, I can now claim that scientific practice is best seen as the collection of actions grouped together (1) because they are geared towards the achievement of the unifying

---

[16] C. S. Peirce, *Collected Papers*, vol. 6, p. 361, (§ 530).

purpose defined above, and (2) because they are performed in ways that involve habits or skills.

Before leaving the topic of scientific practice and moving on to discourse, it is useful to recall that this statement should not be considered a *definition* of scientific practice. In line with the family-resemblance approach adopted earlier, I can hold that one is capable of placing rough boundaries around our idea of scientific practice in so far as one is capable of identifying some collective purpose of all those engaged in it. The notions of practice and purpose go together. The unity of scientific practice is thus not the same as the unity one can talk about between, say, all objects made of gold, which presumably are all composed of a certain kind of atom. On the contrary, the unity of scientific practice is similar to the one we can talk about when considering games.

This is not to say that we have to make do with a lower standard of philosophical analysis. It would be a mistake to insist that the concept of 'game' is a kind of second class concept when compared to those concepts, like 'gold', which allow the formulation of a corresponding set of necessary and sufficient conditions for their correct use in language. If we take language as it is, and not as we might imagine it should be, then there are no first class and second class concepts. The word 'game' functions properly and usefully in language even though games have no one feature in common. One can speak of a 'ludic practice', referring to the behaviour of people while they are involved in games, even though games have no one feature in common. Similarly, the term 'scientific activity' functions properly in language even though scientific activities and skills may have no one feature in common. We can speak of 'scientific practice', referring to the behaviour of people while they are involved in scientific activities, even though such activities may not have no one feature in common. Admittedly, a family resemblance concept is usually more flexible than rigid concepts that can be defined precisely. The extent of the legitimate application of a family resemblance concept can grow over time. In the general statement about scientific practice arrived at above, this aspect of flexibility comes about because of the historical openness the statement includes. If I had defined 'scientific practice' by a comprehensive list of all scientific activities, then the definition would not have applied to any newly encountered activity. But my treatment of the reality of scientific questions allows the possibility that, at a given time, the term 'scientific activity' can legitimately be applied to something to which it has not yet been applied, but which sufficiently resembles activities to which the term has already been generally applied. The list of real questions may change. The rules of methodology may shift.

The obvious objection here is that family resemblance concepts not only do not avoid but even invite sorites paradoxes. If we are content merely with an open statement about scientific practice as the one above, what is to stop us from gradually extending the concept of scientific activity, of which the practice is constituted, to any activity we choose, such as religious ritual? To respond to this objection one should, again, learn from language as it is. One observes that the negation of a family resemblance concept is itself a family resemblance concept. It is legitimate to hold that S is a scientific activity when it sufficiently resembles those things that in the past have been legitimately held to be scientific activities. With the same kind of reasoning, it is legitimate to *deny* that another activity S* is a scientific activity when it sufficiently resembles things that in the past were considered non-scientific activities. There are therefore contrary currents that hold each other in check. A linguistic balance will be maintained by tension since the conditions for asserting that S is a scientific activity and for asserting that S* is not are met simultaneously. This balance occurs even though there are many borderline cases.[17] Not to over-simplify the picture, one should describe the balance between such concepts not in terms of a simple bipolar situation: the tension is not one simply between the concept 'scientific activity' and that of its logical contradictory, 'non-scientific activity'. The tension is normally multi-faceted. It is between the concept 'scientific activity' and various specific contrary concepts like 'ritual', 'game', 'business', and so on.

Consequently, the way I have defended the unity of scientific practice is not essentialist. What I arrived at is a view according to which scientific practice should be considered a unity in the same sense as a thread is a unity, even though the thread is made up of overlapping fibres, no one of which may run through it all.[18] The individual fibres represent genuine groups of activities having common features, having, that is, explicit methodologies in common, explicit real questions in common, and so on. The entire thread may not have common features, but this does not jeopardise its unity. The common features identifiable in scientific activities at one particular moment in the course of history are crucial in the determination of which other activities at later moments will be included within the same category and called scientific.

---

[17] This is suggested in T. Williamson, *Vagueness*, p. 87.

[18] The analogy is from L. Wittgenstein, *Philosophical Investigations*, § 67.

2. Constraints on Scientific Discourse

Can a similar analysis as the one above give support to the claim that it is possible to talk intelligently about scientific discourse in the singular? Just as a language-game, in the sense use by Wittgenstein, is not fully distinguishable from the activity in which it is embedded, so also a discourse. Once it is conceded that talking of scientific practice is possible, then it is apparently also possible to speak of the discourse associated with such a practice. There is obviously a complex combination of discourses associated with the various subsidiary practices included within the one scientific practice. But this complexity does not make the notion of a discourse of scientific practice vacuous. In fact, given the one scientific practice, it is possible to identify a number of different discourses associated with this one practice. To do this, a convenient way is to employ the previously discussed concept of cognitive command.

It will be recalled that cognitive command is attributable to discourses when they involve the possibility of agreement or disagreement: a discourse exhibits a high degree of cognitive command if and only if it is a priori that differences of opinion formulated within the discourse will involve something which may properly be regarded either as a cognitive shortcoming concerning the experiential input or as a case of different access to background theory, or as a combination of both these factors. Using this concept, we can make two distinctions between discourses in general and therefore between discourses of scientific practice in particular.

The first distinction is the obvious one between discourses that exert a high degree of cognitive command and those that do not. If a discourse exerts a high degree of cognitive command, it may be called a *constrained* discourse. As opposed to this, on the other end of the spectrum as it were, if a discourse exerts a low degree of cognitive command it may be called a *free* discourse. Any given practice can have associated with it both free and constrained kinds of discourse. Moreover, a discourse may be part of the practice itself or it may be distinguishable from it. To mention a simple example, one can envisage a discourse associated with the practice of fixing the price of objects by weighing. Over and above, we may have a discourse of the practice of speaking about this practice of pricing by weighing. The former type of discourse may be called internal because it is part of the practice itself and does not objectify what is happening while weighing: in a sense, it is 'using' the practice. The latter type of discourse may be called external because it is distinct from the actual practice which it considers as an object of discussion: the discourse here is not 'using' the practice but 'mentioning' it.[19]

We may have free or constrained, internal or external types of discourse for scientific practice. We have the discourse of scientific practice as we find say in strictly scientific papers involving results and their interpretation. This is an internal discourse. We may also have an external discourse concerning the methodology and sociology of science, or concerning whether science is enjoyable. There is no reason to believe that an internal type of discourse concerning a given practice should always exert the same degree of cognitive command as an external discourse of the same practice. Disagreement about whether clinical biochemistry is enjoyable is not the same kind of disagreement as the one about whether the nutrition of the mother affects the incidence of diabetes in the offspring. The latter is within a discourse exerting a high degree of cognitive command. The former is not. Similarly for the discourses associated with science taken as a whole: disagreement about whether science is boring is not the same kind of disagreement as the one about whether the experimental method is reliable or not. Associated with scientific practice, we can identify at least two kinds of discourse. The former kind of discourse is external and free, while the latter kind is internal and constrained.

The second distinction arises as a reply to the question: what type of constraint is involved in those discourses that exert a high degree of cognitive command? By constraint here I mean preconditions constituting the relatively stable framework within which the discourse is engaged in. Different types of constraints may be identified if we analyse the general situation in terms of cognitive command. There is a certain imperative element inherent in the concept of cognitive command. When a discourse exerts a high degree of cognitive command, its protagonists feel obliged to explain and perhaps eventually solve any disagreement between them by supposing some cognitive shortcoming.

One can envisage different reasons why disputants feel obliged to postulate cognitive shortcoming. Take for example the discourse usually associated with playing chess. This does exert a high degree of cognitive command. If we ask a question like 'what are the possible moves for the black side, given the configuration on the chess-board now?' we take it that disagreement means cognitive shortcoming. One or both disputants are not remembering all the allowed rules of the game, or is not aware of all the pieces on the board. The rules are

---

[19] This distinction between internal and external is similar to the one employed by W. H. Newton-Smith in *The Rationality of Science*, pp. 4-5.

obviously conventional, but this does not affect in the least the importance of the driving force behind the cognitive command. Chess-discourse therefore exerts a high degree of cognitive command but does so because of conventions. Not all cases of constrained discourse are like this. In a discourse concerning simple everyday objects, I can ask the question: 'What happens to this stone if I now let go of it in mid-air?' Here I postulate cognitive shortcoming for a different reason. I do so not because I am referring to conventions but because I know that the way the world is imposes certain limits. Therefore, if this question about the stone is part of a discourse, that discourse will be constrained by constraints that are non-conventional.

Such a distinction within the constrained types of discourse is recognisable also with respect to scientific practice. The way scientific research is made public by being included in reputable journals involves some very strict rules concerning presentation of data, ways of writing reports, acknowledgements, and so on. As aspiring young scientists know, the skill of abiding by these rules is crucial for the development of the work being done in the particular laboratory they work in. This aspect of scientific practice certainly involves a kind of discourse that is constrained. However, the constraints that explain this discourse-characteristic are conventional. They derive from rules set by editors of journals, by manuals of style, and so on. So here we are dealing with a kind of discourse that is constrained conventionally. One of the fundamental claims of scientists, especially if they advocate some kind of realism, is that these conventionally constrained kinds of discourse associated with scientific practice do not exhaust all the kinds of constrained discourse associated with that practice. There is, apart from the conventionally constrained type, another constrained type that is so constrained because of constraints that are non-conventional. It is one thing to discuss how best to write a paper and get it published in a reputable scientific journal, and quite another to discuss the mechanism by which nutrition of the mother may affect the incidence of diabetes in the offspring. The constraints on this latter type of discourse are due either to the kind of world inquirers live in and the kind of beings inquirers are, or to different access to background theory, or to both.[20]

So my discussion up to now has given rise to two important distinctions. First, there are free kinds of discourse and constrained types of discourse. Within the latter, I distinguished between those kinds that are constrained by conventional constraints and those that are constrained by non-conventional constraints. Since it

---

[20] A useful study of how disagreement in science can be due to different access to background theory can be found in A. Lugg, 'Disagreement in Science'.

is the latter kind that has the crucial role of supplying grounds for scientific objectivity, and is thus often simply called *the* discourse of scientific practice, my attention will be turned exclusively onto this area.

3. Experimentation and the Link to Rudimentary Actions

What justification is there for the claim that there are indeed at least some constraints on scientific practice that are not conventional? The justification is constituted essentially by the following chain of implications. Whenever one speaks of scientific practice, one must be implying a procedure based on experimentation. And experimentation necessarily includes some rudimentary actions familiar to everyone in everyday life. Furthermore, such rudimentary actions of everyday life are, indisputably, non-conventionally constrained: they are, as a matter of fact, always within certain boundaries beyond which changes introduced by convention cannot go.

To elaborate each step of this justification, the role of experimentation within scientific practice must be carefully investigated. An experiment can be viewed as an attempt to get a particular process going in relative isolation and to study its detailed workings or record its manifest effects. It needs experimenters who are practically skilled and theoretically informed. That is why theory and experiment are intimately related. Theories within a research programme affect the experimental design, the way the isolation mechanism is controlled, and also the kind of observation recorded, and by what means. Theory plays a role in the way the conditions of human manipulation are perceived, and also in the construction and interpretation of reliable sense-extending and detecting instruments. But we realise that the relation between theory and experiment is even more intimate when we remember how antecedently conjectured theory is crucial even in designating the focus of inquiry in the first place. Moreover, experimentation depends on the assumption that the causal laws identified and controlled under the artificially closed conditions of the laboratory endure and continue to operate outside those conditions in the open world. Here again prior theoretical conjecture is at work in identifying which causes are worth examining and worth isolating from the infinite totality of conditions.

If one accepts this view of what an experiment is, one realises that the manipulation involved in experimentation consists ultimately of simple everyday operations. This is obviously not to say that whatever constrains such simple everyday operations constrains also the sophisticated scientific practice. The fact that the human eye is constrained to detect only up to certain frequencies does not mean that waves of higher frequencies cannot be detected otherwise. The point I want to make is that non-conventional constraints acting on scientific practice are manifest in situations where protagonists of this practice engage synchronically in simple operations and manipulations involving the kind of rudimentary prediction essential for everyday living. However sophisticated the apparatus is, ultraviolet frequencies have to be detected ultimately by means that do not go beyond the constraints of normal conditions, of normal human inquirers: by a turn of a needle, or a smudge on a photographic plate.

Admittedly, not all areas of scientific practice are intertwined synchronically with such simple operations to the same extent. From some points of view, scientific practice may seem to be partially independent of experimentation as described above. This kind of independence started to become more and more significant at a particular period in history, namely the period when protagonists of scientific practice started to specialise: some starting to concentrate more on theoretical aspects, and some more on experimental aspects. It is plausible to consider the 1880s as the period when physics, for example, changes in this way. Before then, those engaged in this discipline were usually people who were just as deeply involved in shaping conceptual understanding as they were in making and using laboratory devices. After the 1880s, physicists began to think of their careers as differentiated into different kinds: theorists and experimental physicists, and so on.[21] When such differentiation occurs, those engaged exclusively in theoretical speculation cannot be considered protagonists of scientific practice if conceived of as an isolated group without significant interaction with other kinds of scientists. In line with the holistic considerations engaged in so far, it is the interaction between what the theorist suggests and what the experimenter confirms that should be the central locus of the legitimisation of scientific truth-claims.

Consider Albert Einstein's bold claim in 1905 that the introduction of the 'luminiferous' ether will prove to be superfluous, implying thereby that the field must be a state of space itself. The claim was not presented as a scientific truth-claim but as an expression of Einstein's belief in the ultimate unity of the physical world and the simplicity of its fundamental processes, and also, obviously, as the conclusion of elaborate and careful mathematical considerations, including a thought-experiment — 'If I pursue a beam of light with the velocity c ...'. The claim came to be considered an official scientific truth-claim only when considered in conjunction with the null result of the Michelson-Morley experiment. But such

---

[21] J. Buchwald, 'Design for Experimenting'.

distance between scientific theory and legitimising experiment was not present in every case. One of the best examples of how scientific practice can be intertwined synchronically with everyday simple operations is certainly that of Michael Faraday's investigations of the nature of electricity and magnetism. The way his speculation was constrained by what the experimental data could allow, and the way this led him to confirm some claims and not others, as seen in his laboratory notebooks, has been the object of careful study by a number of scholars, and, since it lies beyond the scope of my philosophical argument, will not be pursued in any detail here.[22]

These are just a few examples showing how non-conventional constraints on scientific practice become manifest, if at all, within the context of experimentation which involves ultimately the handling of objects in a laboratory just as we do it in everyday life.

Some may find it difficult to hold that the aspect of scientific practice concerned directly with non-conventional constraints is so intimately related to normal ways of acting and behaving. These normal ways are usually relegated to a branch of epistemology altogether separate from the one concerning science. The former is often considered under the title of common sense; the latter under the title of scientific methodology. Moreover, some argue that there cannot be any relationship between the two. W. H. Newton-Smith, for example, has offered an argument that apparently shows that there is a marked difference — a rupture — between common sense and science.[23] One can argue that the former is concerned with noting correlations between observables, while the latter is concerned with suggesting, and subsequently justifying or falsifying, theories that explain these correlations. He admits that common sense can be refined to a high degree because the search for regularities can involve not only observables that need to be more precisely specified, but also correlations that are not so evident and may have no connection with phenomena which concern us in everyday life. But he claims that, even if we incorporate this idea of refined common sense, there is no continuity between it and the scientific method: '[scientific method] is more aptly described as involving a rupture with the procedures of common sense.'

---

[22] D. Gooding, *Experiment and the Making of Meaning*. I am indebted also to N. J. Nersessian, *Faraday to Einstein: Constructing Meaning in Scientific Theories*; R. Harré, *Great Scientific Experiments*.

[23] W. H. Newton-Smith, *The Rationality of Science*, pp. 210-212. He has since then abandoned this view. He argues for the continuity between common sense and science in W. H. Newton-Smith, 'Modest Realism'.

This needs more thought. If we take a closer look at refined common sense and at scientific practice, we can see that the rupture Newton-Smith is trying to identify cannot be considered a definite boundary. We can allude to his own examples. One of the paradigmatic scientific moves he mentions is the one made by Galileo when he postulated the existence of mountains on the moon to explain certain changing patterns as shadows cast by these mountains. This move, Newton-Smith claims, is one which shows that the discovery of correlations between observables is not yet science. 'Science begins when, having noted correlations, we seek an explanation of why they obtain.' Usually, this involves the postulation of entities, like Galileo's mountains, which explain the observed correlations. But is this procedure different in kind from the one we attribute to people exercising common sense? It seems not. Take Newton-Smith's own example of a kind of refined common sense that involves not only the identification of conjunctions of observables but also hunches, as he calls them, that lead to the conjecturing of testable hypotheses: 'Some primitive Popperian man noting that logs floated might have conjectured that a large log would support a man and in putting this to the test invented the boat.' I think it is very difficult to understand what is happening in this example if we were to say that the man was not doing any explanation whatsoever of observed regularities. It is very difficult to understand what is happening without saying that his hunch constituted an explanation: for instance that boats are possible because larger logs carry larger weights, or more mystically perhaps that boats are possible because larger logs are more friendly to humans. Whatever thinking goes on in the Popperian man's mind, his conjecturing of hypotheses is already an attempt at explanation. So from Newton-Smith's standard example of common sense, there appears to be no strict boundary between observation of correlations and explaining them, especially if the one doing the observations uses them for a desired effect. If we go back to his standard example of scientific method, we also see that the boundary is not all that definite. Galileo postulated the existence of mountains on the moon. How could he have done so if it were not because he extrapolated from prior simple observations made under normal, everyday conditions? Wouldn't the Popperian man also have explained, for example, the persistence of a shadow by holding that the dangerous carnivore is still behind the bush? The point I want to make is that the distinction between common sense and scientific method is not a distinct rupture. Although Newton-Smith is right in highlighting the difference between the thought process involved in common sense and that involved in scientific method, it seems better to describe such a difference as a difference in degree rather than a difference in kind.[24]

The main objections that may be brought against this claim seem to be two. The first involves the possibility of having scientists working in our modern sophisticated laboratories solely in the area of correlation-discovery. If such limited work is possible, then it seems that the rupture reappears: correlation-discovery with common sense on one side, and theorising or explaining on the other. But this objection is easily refuted. Having scientists limit their study strictly to correlation-discovery is impossible. Scientists cannot remain on the side of correlation-discovery without in any way sticking their neck out by implicitly suggesting underlying theoretical explanations of such correlations. To see why, consider an example. Take a group intent on optimising agricultural products in a certain region. Their project is, say, to discover correlations between the weather and agricultural output. Can they engage in this correlation-study without a direct or indirect involvement in theorising? The answer is no. Seeking a correlation between the weather and agricultural output involves some crucial preliminary questions. Which aspects of the weather and which aspects of the agricultural output should be taken into consideration? Which aspects could be correlated with which? Should one seek correlations involving atmospheric pressure, or colour of the sky at dawn, or wind speed at ground level? Should one seek correlations involving weight of the agricultural product, or its nutritional value, or its taste? The way the investigators carve up the entities involved so as to make the correlation-study manageable already contains judgements of a theoretical kind bearing on an indirect, tentative explanation. It already contains significant elements of an explanation in the form of a hypothesis that one aspect is more significant than another — that the wind speed, say, is more significant than the colour of the sky. Moreover, such investigators, if honest, should be interested in lawlike rather than accidental generalisations. So here again we see that the work of such scientists cannot be just correlation-discovery totally disconnected from theoretical concerns. If correlation-discovery, because of its apparent simplicity, is associated with common sense, then the necessity of some element of theorising within correlation-discovery confirms what was stated before, namely that the difference between common sense and science is one of degree. There is no boundary which can be defined in any principled way.

---

[24]   The examples treated by Newton-Smith are taken from physics. More striking relationships between common sense and science can be identified in the area of biology: cf. S. Atran, *Cognitive Foundations of Natural History: towards an Anthropology of Science.*

Another possible objection could come from those, usually called constructive empiricists, who hold that scientific claims should be accepted as empirically adequate and not as true. Such a position leads to a distinction between science and common sense because, presumably, constructive empiricists take common sense to consist of statements one accepts as true and not merely as empirically adequate. So, according to this objection, the rupture reappears: science and common sense are different in kind, the former involving empirical adequacy, the latter truth.

But this line of argument is mistaken. One can see, on closer scrutiny, that the kind of rupture described here arises from the idea that, on the one hand, common-sense statements are accepted as true because they deal with observables, while, on the other hand, scientific statements cannot be accepted as true because they often deal with unobservables. The important point here is that, since scientific statements often deal with unobservables, they should allegedly be approached with a different stance: they should be accepted as empirically adequate. But the split between common sense and science does not correspond exactly to the split defined in terms of unobservables. The inference to the existence of unobserved, as distinct from unobservable, entities certainly occurs in common sense. And this inference is no different from that which occurs in science when this deals with unobservable entities. The cognitive process is of the same kind. In both camps, we engage in inference to the best explanation. The inference that leads from the perception of the shadow to the claim that the dangerous carnivore is still behind the bush, is the same kind of inference that leads from the perception of the deflection of cathode rays to the existence of electrons. The inference is of the same kind *even though* the carnivore can, in principle, be seen — if we find enough courage to take a peep — while electrons cannot. So the difference between common sense and science is not a difference in kind but in degree. There is no rupture between common sense and science even though the possibility of making mistaken inferences is less in the former than it is in the latter, where we cannot rely on perception to confirm or falsify claims about unobservables.

Let me now summarise the line of argument in this chapter. My aim was to offer some arguments in favour of the claim that science can be viewed, in some respects, as a single whole. I proceeded by applying the results of the previous chapter. There were three major sections. The first section involved the application of what had been said about *practice* to the case of science, so as to identify the sense in which one can legitimately speak of scientific practice in the singular. An account of the purpose of scientific activity can be formulated as follows: (1) to find explanations of whatever strikes us as being in need of explanation, given a set of real questions; (2) to be critical and weed out unsatisfactory explanations; and

(3) to do this according to a set of methodological rules whose role is not to fix the activity of scientists once and for all but to give indications and foster skill. Once this account of purpose is accepted, one is entitled to hold that scientific practice is best seen as the set of actions grouped together (1) because they are geared towards the achievement of the unifying purpose defined above, and (2) because they are performed in ways that involve habits or skills.

The second section of the chapter involved the application of what had been said about *discourse* to the case of science. The focus of attention was agreement and disagreement in science. Two important distinctions were made. The first distinction was between free kinds of discourse and constrained types of discourse. Within the latter, I identified the second distinction, namely the one between those kinds of discourse that are constrained by conventional constraints and those that are constrained by non-conventional constraints. The latter kind is often simply called *the* discourse of scientific practice. It has the crucial role of supplying grounds for scientific objectivity.

The third and final section of the chapter concerned experimentation wherein both scientific practice and discourse are at work. Some examples were discussed to show how non-conventional constraints on scientific practice and discourse become manifest within the context of experimentation which involves ultimately the handling of objects in a laboratory just as we handle them in everyday life. The claim that there is an important link between science and the handling of objects in everyday life was defended against arguments that could arise if the difference between common sense and scientific method is taken to be a difference in kind rather than a difference in degree.

At this point in the overall argument, therefore, one realises that more light will be thrown on the non-conventional constraints on scientific practice, and on the way judgements are made in this regard, by investigating what exactly is implied by referring to simple observations and actions done under normal everyday conditions, that is by investigating the practice that underpins common sense.

Chapter Six

Common Practice

The best starting point for describing common practice is probably to investigate its links to common sense. Specialised practices, like the practice of dealing with cloud-chambers in all its complexity, include many more actions than are included in common practice. The actions that I am taking to be included under the term 'common practice' are only a small subset of those we usually associate with everyday life for normal people under normal conditions. The collective purpose that can be taken to characterise common practice is the rudimentary goal to survive, a purpose that involves an interest in evaluating sources of information and the corresponding interest in acting according to such evaluation so as to engage in a minimal kind of prediction and a minimal kind of control of the immediate environment.

To define this more precisely, I will proceed by taking two steps. The first involves recourse to the philosophical literature on common sense and on universal rationality. The aim here is to offer some reasons in favour of the claim that common practice is universal. The second section will constitute a refinement of the notion of common practice. It will shed further light on how this kind of practice is situated with respect to a fixed biological nature common to all human beings and with respect to a culture that is not fixed but depends on the local group. In this section, allusion will be made to Wittgenstein's notion of forms of life.

1. Common Practice as Universal

Since common practice underlies common sense, it seems reasonable to attempt to gain some insights into what such common practice must include by taking a closer look at common sense. In epistemology, commonsensism is usually taken to be the view that we know most, if not all, of those things which ordinary people think they know. Moreover, defenders of common sense hold that they know or are justified in believing certain things quite independently of their being able to say how they know them. For example, G. E. Moore maintains that we know that there are material or physical objects having shape and size in three dimensions. His treatment of the question however is primarily directed against a certain kind of scepticism.[1] He pictures himself holding up his two hands and making a certain gesture with the right hand while saying 'Here is one hand' and a certain gesture with the left while saying 'And here is another'. He thereby proves that two human hands exist. According to him, we are dealing here with a proof that qualifies without any problem as rigorous according to the standards set in everyday language. What he wants to show is that common sense should have priority over other ways of thinking — even in telling us what a rigorous proof should be like. In other words the implication 'if sceptical arguments are taken seriously then I cannot know that this is a hand', taken together with the obvious claim that I do know that this is a hand, is a *reductio ad absurdum*. The conclusion is: sceptical arguments should not be taken seriously.

Moore and other common-sense philosophers do not seem to have attempted to define clearer limits on what situations count as common and what situations count as uncommon. Who are these *ordinary* people? It is not at all self-evident that what counts as common sense in one culture counts as common sense in another, that what counts as common sense in one period in history counts as common sense in other periods. We can very well imagine someone, maybe from a social background different from that of Moore, who genuinely feels that holding up one's two hands and making a certain gesture with the right hand while saying 'Here is one hand' and so on, is not common at all. What common-sense epistemologists need therefore is to consolidate their position by showing that it is plausible to suppose that there is something common to all humans engaging in epistemological pursuits. Now, my project is to make a further step and go beyond the epistemological aspects of common sense. I want to employ the notion of common practice: that practice on which common sense is grounded. The consolidation mentioned above is therefore as crucial to me as it is to common-sense epistemologists. I will be assuming that all human societies are *ipso facto* societies of inquirers. I am taking this to be highly plausible as a fixed point to start from because human beings need true beliefs about their surroundings, beliefs that can serve to guide their actions to a successful outcome.[2] However, I am not letting this assumption restrict my view to epistemology: I will not be considering the common elements to be principles dealing exclusively with epistemology. The common elements I will be concentrating on include, in the greater part, the simple handling

---

[1] G. E. Moore, 'Proof of an External World'. See also B. Stroud, *The Significance of Philosophical Scepticism*.

[2] Cf. E. Craig, *Knowledge and the State of Nature*, p. 11.

of objects, ways of moving oneself around, and ways of doing so in a manner that involves a minimal form of prediction teachable from one generation to the next.

So my next task is to clarify the sense in which one can speak of universality as regards common practice. This is *prima facie* a daunting task. Different cultures and traditions have produced innumerable everyday practices which look mutually disjoint. Even such a simple, basic practice as measuring the time of day has had a long and varied history, and not all cultures have held the same priority values as regards what should be considered important. Up till 1873, the Japanese used a system of measuring time involving variable rates, fluctuating hours at day and night, and hours of different length every fortnight.[3] To us, this looks totally impractical. And yet they had no problem. This is just one example of how a certain practice thought to be common to all human societies turns out to be very much dependent on social context. Many other examples can be given. Should this make us give up hope completely? We should not be too quick. Anthropological studies have given firm grounds to believe that all human beings attempt to predict in roughly the same ways. Moreover, any culture that engages in successful prediction must presuppose a given reality common to all. These two points give some preliminary hints about how one should start working towards a justification of the claim that there is such a thing as common practice.

Cogent arguments can be put forward to show that some criteria of rationality are universal. Although these arguments are on the strictly epistemological level, they justify to a certain extent the search for a way to understand a practice that is also universal if such a practice is linked in a special way to these universal criteria of rationality. I will therefore proceed by presenting these epistemological arguments so as to help situate the concept of common practice I want to work with.

Let a criterion of rationality be a rule concerning what would count as a reason for believing something or for acting in one way rather than another. Some philosophers working specifically on theories of anthropological understanding argue, rightly in my opinion, that to make sense at all of the anthropologist's task of reaching out towards and understanding a different culture, of building a bridgehead as they call it, they need to assume that some criteria of rationality are universal.[4] This has to be assumed even though some other criteria are taken to be culture-dependent. Care must be taken to differentiate between two kinds of

---

[3] This is discussed in G. P. Baker, P. M. S. Hacker, *Wittgenstein: Rules, Grammar and Necessity*, pp. 323-324.

[4] Cf. B. R. Wilson (ed.), *Rationality*; M. Hollis and S. Lukes (eds.), *Rationality and Relativism*; D. Lewis, 'Radical Interpretation'.

questions. On the one hand we can ask: when considering a certain society, what are the criteria of rationality in general? On the other hand, we can ask the more specific question: what are the appropriate criteria of rationality to apply to a given class of beliefs *within* the society under investigation? The first question does not allow the answer that the general criteria of rationality are culture-dependent. We cannot even conceive what it could be like for someone to say that these criteria are culture-dependent. As far as the second question is concerned, we are to admit that there are obviously contextually-provided criteria of meaning, and so on. This answer can be given only after the criteria of rationality in general have been assumed to be fixed. Hence the bridgehead may be constructed: the existence of a reality common to both the anthropologist and the culture being studied is a necessary precondition of the understanding of the language of that culture.

But there is a vulnerable spot in the above argument. It could be a contingent matter that all human cultures actually in existence predict in roughly the same way. It could be a contingent matter that all cultures are somehow mutually accessible by employing a hard core of culturally independent criteria of rationality. Nothing has been said about the necessity of having this hard core. Hence to make the justification of universal criteria really strong — to eliminate, for example, the possibility of eventually discovering a new tribe whose criteria of rationality are totally different from our own — we have to engage in some kind of *reductio* argument. The aim is to show that assuming the existence of a group of beings who think and reason in accordance with thought-constraints completely different from our own leads to an absurdity. For this purpose, I can allude to Donald Davidson's well-known argument.[5] Think of what is involved in this imaginary encounter with such beings. We are presupposing two steps. First we make the judgement that they are capable of thinking or reasoning. Then we conclude that their thinking and reasoning is completely different from our own. But here the inherent problem comes to light. At the first step, how can we identify a process and call it thinking if it is completely different from our own? For thinking to be recognised as thinking at all, it must have some common element with *our* thinking. The constraints we have on thinking partly define what counts as thinking.

Another way of making the same point is to refer to the well-known quotation from Quine: any statement can be held true if we make sufficiently drastic changes elsewhere in the system. In this case, therefore an objector may insist that we can indeed accept that these beings we meet are in fact thinking even though such a process is completely different from what we call thinking. To accept this, all we

---

[5] D. Davidson, 'On the very Idea of a Conceptual Scheme'.

need is to make the right adjustments within our presuppositions. However, the case under consideration is not so simple. It amounts to abandoning what we call thinking. To accept that these beings are thinking completely differently from us, we need to make not only drastic changes *within* the system — that would not be enough. We would also have to renounce one of the main points of having such a system at all. We would have to renounce the primordial attempt to clarify the world to ourselves in any intellectually satisfying way. And if *that* is allowed, then anything goes.[6] So in the end we always arrive at an absurdity. The *reductio* argument is complete. We are entitled therefore to hold that there is necessarily a hard core of culturally independent criteria of rationality.

An objection here has to be faced. In the above treatment of the problem, only bipolar situations were considered: our culture and their culture; our thinking and their thinking. If we introduce a slightly more complicated situation, which will bring the entire discussion closer to the actual world we live in, then a special problem arises. Given three societies, A, B and C, for a member of A to understand a member of C there must be a core of principles of rationality common to both. For a member of B to understand a member of C, there must likewise be a core of principles common to both. But there is no reason to suppose that the same core of principles at work is the same in both cases of interaction. What is necessarily kept constant in one interaction may be completely different from what is necessarily kept constant in the other. Hence one is not justified in working with the assumption that there is a core of principles common to *all* human inquirers, even though it is legitimate to hold that there is a core common to any two groups.[7]

A response to this objection may be given by running again at a more fundamental level the Davidson-type of necessity argument: this time not in terms of 'thinking' but in terms of 'being human'. Davidson argues that, for thinking to be recognised as thinking at all, it must have some common element with *our*

---

[6] The point about the quotation from Quine is made in E. J. Craig, 'The Problem of Necessary Truth'. Fundamental premises associated with thinking have had various formulations. Kant calls it 'the transcendental unity of apperception', signifying thereby the fact that all of the representations I have are necessarily mine. For Strawson, the essential point is that it must be possible for someone to distinguish between states of himself and states not of himself. For Ross Harrison, in any comprehensible world there must be reasons available to the protagonist enabling him to distinguish between those judgements of his which are true and those judgements of his which are false. A valuable source of information about these issues is R. Harrison, *On what there must be*.

[7] Cf. D. Lewis, 'Radical Interpretation'. He attempts to give the complete set of questions to be answered when dealing with the problem of radical interpretation but he does not include this problem because he never explores beyond the context of a bipolar situation.

thinking. Here I will argue that, for human societies to be classifiable as human, they must have some common movements, actions, practices. The objection mistakenly takes the interaction between societies to be an interaction between propositions: as if a member of one society arrives with his bag of accepted propositions and faces a member of the other society who arrives with her bag of accepted propositions, the question being which propositions are to be kept and which discarded when some of one bag contradict some of the other bag. But this view is made up exclusively of intellectual and discursive considerations. The objection does not allude to simple kinds of movements and practices human beings engage in, given their biological characteristics, which are significantly similar, and given their surroundings, which again in a number of fundamental respects are also similar. If we concentrate on these aspects related to what humans do, then we realise that, once we start with the proviso that what is under investigation is interaction between various *human* societies, then we have already assumed a minimal universality between all the members of A, B and C.

My extension of Davidson's argument could be seen as an extension of his Principle of Charity, a weak version of which is: If we want to understand others, we should count them as believing what they ought to believe, given a certain degree of similarity between us and them. The extension from 'thinking' to 'being human' turns our attention to the fact that, when we are faced with the problem of radical interpretation, what is fundamental is our theory of persons. The problem must start here: it must start with a theory of persons, a theory which is probably implicit. This theory ensures that the basic concepts of belief, desire and meaning are common property for us and for all groups of interlocutors. David Lewis attempted to improve on Davidson's Principle of Charity on these lines, and his suggestions add support to what I am arguing for.[8] According to his improved principle, we must hold that there is an inductive method that holds for us and also for our interlocutors, as far as we can see. One is not dealing with propositions anymore, but with the grounds justifying relations between them. The beliefs of our interlocutors and our beliefs are related as follows. Our interlocutors have access to a certain kind of evidence, which Lewis calls their life history of evidence, which we may call E. From our standpoint, we will not have a complete view of E. So let us call E* the life history of evidence as we see it. The same kind of distinction can be made between the beliefs they actually have, B, and the beliefs we attribute to them, B*. Now Lewis is making the valuable point that the inductive method we

---

[8] D. Lewis, 'Radical Interpretation'. Lewis does not make the distinction I introduced in chapter three §2 between belief and acceptance. His account may be taken to involve *accepted* beliefs.

use to go approximately from E* to B* must be the same as the method we use to go approximately from *our own* pool of evidence to *our own* beliefs. Lewis reminds us that the qualification 'approximately' is important: our common-sense theory of persons tells us that the beliefs and desires of our interlocutors may differ from ours not just because of the different evidence available to our interlocutors but also because of the effects of the interlocutors' different training. But this should not make us abandon the fundamental idea of having a theory of persons to start with. If our common-sense theory of persons told us just what these effects of training were, we could build them into a still better version of the Principle of Charity.

So the line of argument justifying the existence of a universal core of rationality leads to the conclusion that some kind of practice must also be common to all human societies. This common practice will presumably include the simple elements of behaviour that are intimately linked to the fact that human inquirers are the beings they in fact are, and intimately linked to the fact that the world they operate in is the way it is. As a first approximation, we can say that this common practice probably includes handling things, moving around, and also the rudimentary kind of predicting that is needed to do so.

2. Common Practice Situated between Biology and Culture

To further refine the concept of common practice, I will now situate it with respect to the related Wittgensteinian concept of 'forms of life'. Although I will not be going into any exegetical details concerning the exact meaning Wittgenstein had in mind when he used this term, I will nevertheless mention, and subsequently enlarge upon, some possible interpretations that have been discussed in the literature.[9]

Of the three interpretations to be mentioned, the first will not be dealt with at any length. It involves the identification of forms of life with language-games. On this view there are as many forms of life as there are language games. But it is difficult to see how this can account for all of Wittgenstein's comments. Some uses of the term suggest that language-games are embedded in a form of life. According

---

[9] See J. F. M. Hunter, '"Forms of Life" in Wittgenstein's Philosophical Investigations'; G. P. Baker and P. M. S. Hacker, *Wittgenstein: Rules, Grammar and Necessity*, pp. 229-251. It is interesting to notice that 'form of life' is used only five times in the *Philosophical Investigations*: §§ 19, 23, 241; pp. 174, 226. He alludes to the same concept in other writings, e.g. *On Certainty*, §§ 356, 358, 359.

to a second interpretation, a form of life is a kind of set of interrelated tendencies to behave in various ways: to have certain facial expression and make certain gestures, to do certain things like count, help others, and so on. The interrelation between these behaviours arises because a subset of them will be associated with a certain action. For example pitying someone means having *that* particular subset of kinds of behaviour and gestures. Likewise for other actions. This view implies that by 'form of life' Wittgenstein meant something very close to what I previously called common practice. However, given my added strong condition that common practice is universal, the interpretation that is most interesting is the third I want to mention, namely the one favoured by J. F. M. Hunter. According to this view, a form of life is something typical of a living human being. By 'typical' here is meant that which characterises human beings with respect to other kinds of beings. Hence we are surely including the biological vital functions differentiating living from non-living things. But more has to be included. Hunter includes those aspects of which we are *not* directly conscious. These involve the normal biological make-up, the unlearned elements of our physical functioning, and also other aspects which, though learned sometime in the past, are now carried out automatically:

> we can move by easy stages from automatic, unwilled, not-conscious processes like nutrition, through reflex actions, many of which are learned or at least *acquired*, and which, though not done at will, can often be *resisted* at will; then through speaking or writing just insofar as it is forming the words with our mouths or drawing the characters on paper, where, though we may form a word at will, we do not (generally) will the physical manner of our forming it; to, finally, expressing ourselves *in a certain way*, where although it is generally done at will, we do not will the *willing* of it, and we do not know how just this form of words satisfies all the various grammatical, social, personal, and intellectual requirements of being something we 'want to say'. We may by studying it afterwards *find out* how it satisfies such requirements, but the interesting thing is that we generally manage to say things which are just about what we *would* say *if* we had the requirements in mind, but *without a thought* of the requirements.[10]

This third view is confirmed by the exegetical work of G. P. Baker and P. M. S. Hacker. According to these, Wittgenstein did not have a concept of 'form of life' which is predominantly biological as if the only use he wanted it for was to differentiate a human form of life from a Martian one, presumably based on a different biological nature. For Wittgenstein, common elemental human practices,

---

[10] J. F. M. Hunter, '"Forms of Life" in Wittgenstein's Philosophical Investigations', pp. 278-279.

the engaging in which makes us human, can be called 'facts of human natural history'. But here again, what is 'natural' is not uniformly biological. Looking in the direction pointed at rather than at the pointing finger, crying out in pain, laughing when amused, and so on, are *biologically* natural. Continuing the series of natural numbers '1001, 1002, 1003, ...' is *culturally* natural. This latter is natural for us, but not for all people at all times and places. For Wittgenstein, human natural history is predominantly anthropological: 'commanding, questioning, recounting, chatting, are as much part of our natural history as walking, eating, drinking, playing.'[11] Given this pluralistic notion of 'natural', we can see that Wittgenstein's notion of human nature is not predominantly biological. This point makes Baker and Hacker insist that the idea of 'form of life' is to be clearly distinguished from the common biology of human beings: 'Wittgenstein's conception of human nature is not predominantly a biological one, then *a fortiori* his concept of a form of life is not biological, but cultural.'[12] The point they are making here seems to be that a form of life is not exclusively biological. Even though, just like culture, one cannot have a form of life without a biological substratum, Wittgenstein's attention was not on this universal biological substratum.

A question may arise whether we can ever pretend to be able to stand back, as it were, and survey our form of life, given that we are necessarily embedded within it. The very act of considering *our* form of life, if at all possible, would itself be part of our form of life. Jonathan Lear suggests that Wittgenstein should have distinguished between the 'we' and a 'form of life'.[13] The latter corresponds to Kant's 'I think', which can be predicated of all our judgements. Likewise, 'form of life' is a predicate which may be predicated of various objects. We may use the term narrowly and label disparate social groups alternative 'forms of life', or we may use the term widely to mark the form of life which we all constitute. In this latter sense, however, it is not so clear what we are talking about by using the plural pronoun 'we'.

It is arguable that the relation in which we stand to ourselves is similar to the relation in which I stand to myself. Now, every representation of the self must fail to capture fully the self as it is in itself, because that consciousness which must be able to accompany each of my representations cannot itself be adequately represented by any of them. In fact, there are cogent reasons to hold that the 'I' is

---

11 *Philosophical Investigations*, § 25.

12 G. P. Baker and P. M. S. Hacker, *Wittgenstein: Rules, Grammar and Necessity*, p. 241.

13 J. Lear, 'Transcendental Anthropology'.

not a name or another kind of expression whose logical role is to make a reference.[14] Similarly if we switch to the first person plural: 'form of life' in the universal sense is a reflective concept, used to construct a representation of us. We construct a conception of ourselves as acting in similar ways on the basis of shared interests, beliefs and desires. But again, some aspect of our subjectivity must be left out of this representation. If the 'we' encompasses us all, encompasses any being who might in the widest sense count as one of us, then, to arrive at any content of this notion, we are apparently obliged to remove each and every feature that is linked to a local form of life. This inevitably leads to a situation in which we are left with nothing but a vacuous metaphysical subject.

Does this mean that nothing can be said about the form of life that encompasses all of us? Not necessarily. There is the possibility of making true claims about us which, even though they are expressed in the anthropological vocabulary of needs and desires, are not confined to one particular form of life. It is certainly conceivable to make claims meant to express the conditions of possibility of being an inquirer in any way at all, of being engaged in a form of life at all. Further details on how such transcendental considerations may be engaged in to arrive at general constraints will be discussed in the next chapter. At this stage, it is enough to secure the point that such claims are indeed possible even from within our form of life.[15]

Now, if we accept this interpretation of Wittgenstein's concept of form of life, how is the previously elaborated notion of common practice situated with respect to it? I will make two points. The first one is that, when discussing notions like human nature and forms of life, in the sense used by Wittgenstein, we are evidently dealing with a broad kind of notion which can, and often does, cover at least two opposing ideas. At one extreme we can be referring to a biological substratum common to all inquirers like us, because it establishes what 'us' means in a minimal way by reference to simple needs — such as the need to say something meaningful. At the other extreme as it were, we can be referring to 'culturally natural' aspects, in the sense described above: aspects which are different according to time and place. The common, fixed biologically natural aspects allow a multiplicity of culturally natural aspects. It may seem for some readers that Baker and Hacker are suggesting, rather briskly, that Wittgenstein's notion of human nature contains both extreme ideas, while his notion of form of life contains only the culturally natural aspects. But this temptation, it seems to me, should be resisted. Although Wittgenstein's attention

---

[14] G. E. M. Anscombe, 'The First Person'.

[15] Cf. B. Williams, 'Wittgenstein and Idealism'; J. Lear, 'The Disappearing "We"'.

was not on the biological substratum, he seemed to be including *all* the limitations
— arising biologically or otherwise — of beings that are human.[16]

The second point is that my notion of common practice, when considered in
these terms, is situated at the interface between these two poles. Even though I want
to include a biological nature common to all humans, which I am assuming to exist,
I am not limiting the concept of common practice to this substratum. I want to
venture out, as it were, to the first layers of discourse, expression, and meaningful
action allowed by these universal biological features. This practice, involving
discourse and repeatable actions with simple predictive success, is so rudimentary
as to retain the universality in time and space attributable to the biological aspects.
Talking about such kinds of practice is possible because we do not have any reason
to believe that the interface between one extreme and the other is a definite dividing
line. Nothing indicates that all human aspects are either clearly biological and
universal, or clearly cultural and local. There is a grey area between what is
certainly biological, like the number of chromosomes, and what is certainly
culturally local, like style of dress.

To conclude therefore, I have argued in favour of two main points: first, that the
claim that there is a practice which is universal is plausible because the
corresponding claim that a core of criteria of rationality is universal is plausible;
second, that this common practice covers part of the area between what is culturally
natural, which is surely local, and what is biologically natural, which is surely
universal.

The role of biology in the elaboration of the concept of common practice,
however, together with the role of other empirical judgements, must be handled
with care. My ultimate aim is to give an account of the objectivity needed for a
realist understanding of scientific practice. Apparently, I cannot resort to the details
of human biology to show that there is something universal on which one can base
such an account. That would be an account dependent on one particular science and
thus it would seem to beg the question.

The reasoning here, however, although it could be labelled circular, is not
viciously so. I have attempted a *philosophical* argument to show that there is such a
thing as common practice. By 'philosophical argument' here I do not mean an

---

16   He includes for example such remarks as: 'our language-game is an extension of
primitive behaviour', *Zettel* § 545. When considering the most challenging case, namely that
of logic, he argues that philosophy's task is to remind us that logic itself is the outcome of
reflection on our practices of arguing rigorously, but that does not mean that anything goes:
'One might say: the axis of reference of our examination must be rotated, but about the fixed
point of our real need', *Philosophical Investigations*, § 108.

argument from within a distinctive and isolated realm of reflection: a realm of reflection that is strictly 'non-revisionary'.[17] I am not taking philosophy to be a self-contained activity providing insight into our other activities which are themselves unaffected by philosophical reflection, and which do not impinge back on the nature of philosophy itself. If one accepts a dialectical interaction between realms of thought, as I am doing here, then the universality of biological human nature, drawn from empirical inquiry, adds support to the claim that common practice is universal, but it adds this support without making this claim useless as a basis from which an account of scientific objectivity can be given. The circularity is not vicious. It just manifests one possible dialectical interaction between two realms of inquiry: the empirical and the philosophical.

More specifically, one claim that the empirical investigation is supporting — to start off the non-vicious circularity, as it were — is the claim that all humans normally have the goal or purpose to survive, in terms of which a particular practice is conceivable in the way investigated in chapter four, §2.

But if intention is universal, what one intends may not be. If speech is universal, what one says may not be. My claim is not the naive one that *to engage* in some practice is universal. That is easily acceptable. My claim is that common practice is universal in the sense that movement, intention, and speech are universal, *together with* some constraints on these possibilities of movement, intention and speech. The constraints here are those that depend on what the agents are, and on what the world they are in is like.

---

17 Wittgenstein is successful in arguing in favour of the non-revisionary nature of philosophy only when he considers cases of non-self-conscious practices. But, since self-reflection occurs in the majority of practices, his point is not very convincing. This observation is made in J. Lear, 'Transcendental Anthropology'.

Chapter Seven

Constraints and Objectivity

In this chapter the aim is to determine how objectivity may be understood in terms of the constraints on a given practice. There will be three main sections. First, I will determine the sense in which we can claim that some constraints on a given practice are non-conventional. This term has been employed before, but in the preceding text it was mentioned only briefly as regards scientific practice, while now a more general approach will be engaged in. Allusion to two relevant aspects from Wittgenstein will constitute the preliminary ground I start from. In the second section, I will make two qualifications to these Wittgensteinian insights so as to arrive at a plausible account of a non-conventional constraint on a given practice. The final section will involve the application of this plausible understanding of non-conventional constraint to our understanding of objectivity.

1. Preliminary Ideas about Constraints

The first point from Wittgenstein I want to discuss is that natural language-games are conditioned by facts. Such ideas concerning conditioning can throw light on my use of the term 'constraints'. Take the example discussed regarding the language-game of buying and selling with a price corresponding to the weight of things.

> It is only in normal cases that the use of a word is clearly prescribed; we know, are in no doubt, what to say in this or that case. The more abnormal the case, the more doubtful it becomes what we are to say. And if things were quite different from what they actually are — if there were for instance no characteristic expression of pain, of fear, or joy; if rule became exception and exception rule; or if both became phenomena of roughly equal frequency — this would make our normal language-games lose their point. — The procedure of putting a lump of cheese on a balance and fixing the price by the turn of

the scale would lose its point if it frequently happened for such lumps to suddenly grow or shrink for no obvious reason.[1]

In this practice of fixing the price by weighing, and in the language-game we associate with it, we are admittedly dealing with something conventional as far as the units of measurement and the apparatus used are concerned. However, there is an essential necessary condition behind both the practice and the language we associate with it, namely the fact that our world is such that objects do not change their weight unexpectedly and for no reason. Examples like this abound in many contexts. In soccer, no matter how radical we want our changing of rules to be, we cannot have a new rule saying that no player is allowed to kick the ball in such a way that it remains suspended in mid-air. Such a rule does not make sense in the type of world we live in — precisely because of the constraints imposed by this world. There are non-conventional aspects of language-games as there are for soccer. These aspects are rooted in pre-linguistic behaviour. As Wittgenstein explicitly remarks, 'it is our *acting* which lies at the bottom of the language-game'.[2] In other words, some specific language-games are responsible to non-conventional facts in the sense that if the world were different, certain features of them would no longer be useful.[3]

So this is one way we can start to understand non-conventional constraints on a practice. Another relevant point from Wittgenstein concerns the propositions of which we are certain in a given situation. Think of a clinic.[4] I go to the doctor, show my injured hand. I do not start my conversation by saying 'This is a hand ..., I've injured it, ...'. The first part of the sentence is completely inappropriate because if, for the doctor, it is open to doubt whether that is a hand then many other well-held beliefs will also be open to doubt, even whether, say, I am human. In that

---

[1] L. Wittgenstein, *Philosophical Investigations* § 142. Other places where Wittgenstein discusses the same point include: *Philosophical Investigations* § 174; *Zettel* § 540-541. For the ideas I develop here I am indebted to a certain extent to D. L. Phillips, *Wittgenstein and Scientific Knowledge*, chapter 4; G. P. Baker and P. M. S. Hacker, *Wittgenstein: Rules, Grammar and Necessity*, chapters 5 and 6; P. M. S. Hacker, *Insight and Illusion*, chapter 7.

[2] L. Wittgenstein, *On Certainty* § 204.

[3] In the following paragraphs, I will be elaborating an argument I sketched in L. Caruana, 'Beyond the Internal Realist's Conceptual Scheme'. For valuable insights into the relationship between necessity and usefulness in Wittgenstein I am indebted to G. P. Baker and P. M. S. Hacker, *Wittgenstein: Rules, Grammar and Necessity*, p. 330. Lars Hertzberg develops a similar argument to the one I am making here in the context of the problem of scepticism. See his 'On the Factual Dependence of the Language-Game'.

[4] L. Wittgenstein, *On Certainty*, §§ 460, 461.

particular context, the question whether what I am showing him is a hand or not simply does not arise. Wittgenstein explains the same point by using the image of hinges on which a door turns:

> [T]he *questions* we raise and our *doubts* depend on the fact that some propositions are exempt from doubt, are as it were like hinges on which those turn. That is to say, it belongs to the logic of our scientific investigations that certain things are *in deed* not doubted.[5]

So to raise the question whether I am certain or not about a proposition is already to cease using it as a hinge. Wittgenstein's use of certainties here is similar to my use of constraints on the practice or discourse engaged in. The hinge, or that which is exempt from doubt within the practice, constitutes some of the constraints on that practice. We see through them, as it were. We engage in the discourse by not paying attention to them.

2. Refinement of the Idea of Constraint

Wittgenstein's insights suggest that constraints on a practice can be thought of as either the normal conditions under which the practice is engaged in, or the propositions which are accepted as a precondition so as to enable the practice to be engaged in. These two points clarify what a constraint is, but they do little to distinguish between different types of constraint. Since my aim is to arrive at an understanding of a constraint of a certain type, namely a non-conventional constraint, some qualifications have to be made to these Wittgensteinian insights. I will make two: one dealing with the idea of hinges and the other with the descriptions protagonists of a language-game give of the constraints within that language-game.

Hinges

Can anything more be said about the *type* of hinge we are keeping constant? When discussing the clinic, I situated the discussion in a particular context, and this allusion to the context within which the discourse is held is important. My visit to the doctor as described above is presumably taking place in normal conditions: this

---

[5] L. Wittgenstein, *On Certainty*, §§ 341, 342.

is what makes the claim 'this is a hand' so inappropriate. But we can imagine a situation where it would not be inappropriate: for example in trench warfare where doctor and patient are in the dark, where injuries can be terrible, and so on. What used to be a fixed hinge in one context has ceased to be so in another. This shows that it is wrong to assume that whatever we are certain of, whatever we cannot doubt in a particular context, is true. One should not identify certainty with truth.

But what is a context? Different views on the nature of a context will inevitably result in different views on what Wittgensteinian hinges are. It would be a mistake to assume that the context involves only the physical surroundings of the two protagonists. The examples above began with the statement that patient and doctor are both in a clinic, or both in a trench. What this is hinting at is that the context is meant to be binding on both protagonists. But is it binding on them in the same way? This is not so clear. How can Wittgenstein be so sure that what counts as a hinge for one person is also considered a hinge for the other? This question, which is apparently not tackled by Wittgenstein, derives from the observation that hinges are not dependent only on the physical make-up of the context — the clinic as a particular room, or the trench as a collection of particular physical conditions — but also on the way this physical make-up is perceived from the particular point of view of each protagonist.

But, if considerations involving points of view are introduced, as they inevitably should, we may be drawn into a argument leading to the conclusion that agreement, in the strict sense of the word, is impossible. One protagonist has a point of view governed by some presuppositions, while the other has a point of view governed by some other presuppositions. Even though the presuppositions are not known explicitly, the fact that they do not coincide completely may give the impression that whatever agreement the protagonists arrive at is only apparent. At best, it is very shallow — only a few presuppositions deep, as it were. If one were to ask more and more questions about the protagonists' way of acting, then it will turn out that the hinges for one are not the hinges for the other.

The problem with this type of reasoning is not that it is expanding the idea of context beyond the physical surroundings: as if avoidance of considerations involving points of view could enable us to hook our understanding onto the physical aspects of the context and thus secure a common hinge between the two protagonists. The problem arises because this type of reasoning does not expand the idea of context far enough. To take seriously all the implications of the notion of language-game, we have to hold that the context includes not only the physical surroundings together with the presuppositions particular to different points of view, but also the fact that communication is successful and genuine agreement possible. This is the starting point. The idea of hinge therefore is not attributable to one single individual protagonist. That would make the idea of hinge and the idea of a set of presuppositions identical.

From these considerations therefore, one can conclude that Wittgenstein's hinges can change from context to context, context here being understood as including more that just the physical surroundings and the protagonists' sets of presuppositions. If the non-conventional constraints on a given practice are to be described in terms of Wittgensteinian hinges, then they must be described as hinges that are context-independent or universal. When the practice under investigation is common practice, it is not difficult to conceive of non-conventional constraints of this kind, given that common practice was defined as that practice which is universal because it lies very close to biological nature.

Descriptions

The second qualification concerns the descriptions we can give to these constraints. When discussing the practice concerning fixing the price of cheese by weighing, it was mentioned that the practice would lose its point if it frequently happened that lumps of cheese suddenly grew or shrunk for no obvious reason. That is incontestable. But, the issue certainly involves what scientists call, or used to call, the law of conservation of mass. What Wittgenstein was suggesting by his example corresponds to what scientists present in the form of a law, supposedly describing one of the fundamental characteristics of the world. Moreover, it is undeniable that this law has suffered a lot of changes. In fact, even the very concept of mass has undergone considerable refinements since antiquity.[6] And because of recent developments, as is well known, we do not speak anymore of the conservation of mass but of the conservation of mass-energy. What used to be considered strictly unchanging has been shown to be changeable after all. If history teaches us that it is risky to claim that scientific laws are unchangeable through the ages, then it is wrong to try to extract a fact about the real world from the Wittgensteinian example of the practice of weighing.[7]

But this objection can be resisted. Its weak point is that it does not do justice to the kind of relationship there is between everyday practice and scientific practice. It is true that nowadays we do not consider the law of conservation of mass as

---

[6] Cf. M. Jammer, *Concepts of Mass*.

[7] This is a variation of the problem of the Pessimistic Induction. Cf. L. Laudan, 'A Confutation of Convergent Realism'.

obtaining. But this is the case only as far as high-energy physics is concerned. In everyday life it still holds. The disappearance of our language-game of weighing will come about if the law of conservation of mass does not hold *in our everyday practice*. And, as we know, this is not the case. In other words, everyday violation would certainly make the language-game impossible, but high-energy violation does not. So the question is: what role is played by the description of a precondition inscribed in this practice? The description of the constraint tells us that, *for everyday life,* lumps of cheese do not suddenly grow or shrink for no obvious reason. And this is something that relativity physics not only does not refute, but also seeks to be in accordance with — to save face, as it were. The practice of fixing the price by weighing tells us something which all future scientific theories are bound to consider a constraint.

What is being alluded to here should be distinguished from the usual problem of reduction of one theory by another. The question of reduction is usually taken to consist of the derivability of one theory from the other. Hence, for example, it has been argued that physical optics is fully reducible to electromagnetic theory because the former can be shown to be a consequence of the latter.[8] But this approach is applicable only to a very small number of theories. When change of meaning of central theoretical terms is taken into consideration, one has to concede that a successor theory is most likely incapable of either contradicting or confirming all the claims of the old theory. There is no Hempelian-type of deduction from one to another. But this does not preclude the predecessor and successor theories from standing in some more elaborate formal relationships. That the laws of the successor incorporate a close approximation to that of its predecessor is indeed one such formal relationship. The condition of possibility for a theory to be a successor to another is that it be capable of explaining the apparent success of the predecessor laws. The successor theory must, if the case arises, be able to explain why some central concepts of the old theory do not refer and also why some of its concepts are incommensurable with the corresponding ones of the old theory.

This account of reduction is however deficient at least on one important point: it does little to include the context within which the two theories are considered to be applicable. Theories are not isolatable from other theories. They should not be considered as independent from the type of observations they are meant to explain. Some allusion must therefore be made to at least the following three facts: firstly to

---

[8] R. M. Yoshida, *Reduction in the Physical Sciences*, pp. 32-35. For my views on reduction in general, I am also indebted to J. Rosenberg, *One World and our Knowledge of it*; W. Sellars, *Science Perception and Reality*.

the fact that a given theory is held together with a host of other theories, secondly to the fact that a theory is proposed with respect to a certain set of measurement procedures which may be available in one period of history and unavailable in another, and thirdly to the fact that a theory is proposed with respect to a particular range of observation systems that may be accessible in one period in history but inaccessible in another. Jardine has proposed the following version of reduction to include these points:

> Let T be the set of theories held by protagonists of theory T; let M' be the measurement procedures available to protagonists of theory T'; and let K' be the range of objects and systems accessible to protagonists of T'. A first shot at an account of reduction is as follows:
>
> T' is reducible to T if, and only if, there is a theory T^* derivable from T (if necessary with the help of boundary conditions, bridge laws, auxiliary hypotheses, etc.), such that T^* is evidentially equivalent to T' given T, M', and K'.[9]

Does this account of reduction throw any light on the relationship between, on the one hand, conditions of possibility of practices and, on the other hand, subsequent scientific theories that apparently violate these conditions?

One constraint on the practice of fixing the price by weighing was described as the fact that lumps of cheese do not grow or shrink without reason. If this fact does not hold, the practice is not possible. We may be tempted to understand this observation about constraints by holding that what is involved here is a kind of simple rudimentary theorising procedure. We have a theory about the world derived from simple observations of the practice. With our recent theories however we know that if we use a microscope, we will certainly notice that lumps of cheese sometimes shrink, perhaps by sublimation, and they sometimes increase in weight, perhaps by absorbing water vapour or by fermentation. So, if the range of observation systems and the measurement procedures differ from the everyday case, what was considered a constant is now seen to be variable.

When we say that the practice of fixing the price of cheese by weighing is constrained by the fact that such lumps do not shrink or grow for no obvious reason, we are using the words 'shrink' and 'grow' in a specific way. These words get their meaning from the perceptive capabilities of humans who engage in the practice under normal conditions. If all humans somehow change overnight and start having microscopic vision and extreme sensitivity as regards changes in

---

[9] N. Jardine, *Fortunes of Inquiry*, p. 62.

weight, then our present theories tell us that the practice of fixing the price of cheese by weighing will be impossible.

Hence my point is not the same as the one made by those who argue in favour of a strictly empirical basis for all future theories. What I want to emphasise is the fact that, for perceptions to be the basis of a simple theory which is binding on all future theories, they have to be perceptions of normal people under normal conditions. Perceptions are therefore useless on their own. The entire combination of perceptions, bodily movements, intentions and discourse, constituting the practice, is what lies at the bottom of the theory-reduction chain.

A constraint on a practice does not tell us anything for certain about the external world. It does not fix one of the parameters that all future scientific theories will be obliged to hold as fixed. The fixed feature is the dependence of the practice on the constraint. It is this dependence that is fixed for all future theories.

The descriptions of constraints, however, like all other descriptions, can obviously be good or bad, accurate or inaccurate, in one language or in another, and so on. Saying that, through an investigation of the practice of weighing, one may arrive at the conclusion that lumps of cheese do not change their weight for no obvious reason is already allowing some scientific theoretical notions to creep in. 'Weight' as such is already a theory-laden term. And once theory comes in, then the objection that present scientific theories may turn out in the future to be false gets a grip. So although I can say that there are constraints, it is apparently very difficult to say exactly what they are. One has to employ something which is not theory-laden. One has to peel off, as it were, the theory-laden layers surrounding the notion of weight to arrive at more primitive concepts like, for example, that of 'heft'. But then again, if one insists that the meaning of any notion is always to be grasped within a context, even 'heft' would be considered theory-laden to a certain extent. So how can one understand our knowledge of constraints, if we do have any?

One way is to hold that, if we are skilful in the given practice, then we have tacit knowledge of its constraints. An objector may claim that this way of understanding is insufficient. Constraints have a definite effect on the discourse associated with a given practice: they explain how the discourse exerts a high degree of cognitive command. But now, if tacit knowledge is presumably so elusive that it cannot be fully rendered into formal statements, is this tacit knowledge ever enough to ensure cognitive command within the discourse?

This kind of worry is based on a misunderstanding of what cognitive command is. A discourse is one thing; what a protagonist of that discourse knows is another. Whether a discourse exerts a high degree of cognitive command or not, does not depend on any one protagonist's knowledge. Suppose I engage in a discourse concerning hair-styles in rural Japanese society. It is very probable that I start off by considering such a discourse as exerting a very low degree of cognitive command, because in Western societies different opinions on the significance of a given hair-style are tolerated without any problem whatsoever. But then I discover that for rural Japan such a discourse is governed by strict rules, concerning age, marital status, and so on. This illustrates how one can participate in a discourse one thinks is non-constrained, and find out, after interaction with other protagonists, that it is in fact constrained. Such situations show that a discourse has characteristics that do not depend on any one particular protagonist. Whether I know discursively anything about the constraints or not, the discourse still exerts a high degree of cognitive command. Hence, it is not self-contradictory to claim both that constraints explain cognitive command, and that our knowledge of these constraints is often tacit. Engaging in a practice does not directly involve the ability to describe the constraints within which it is played, any more than mastery of tennis requires the ability to specify the laws of physics.[10] It may be difficult for us to see the constraints, but such a difficulty should not be taken to mean that only the superficial aspects of the practice deserve our attention.[11]

There seem to be two related ways of understanding the reliability of a given description of a non-conventional constraint. The first depends on the observation that practices can undergo variations. Fixing the price of cheese by weighing can, and has been, carried out using various types of weighing techniques, using various types of money, using various types of languages, and so on. These variations of the practice however have always kept at least one constraint fixed, namely the one tentatively described by the phrase: 'lumps of cheese do not change their weight for no obvious reason'. For that particular practice, therefore, this phrase qualifies as a good description of one of the non-conventional constraints acting on it. The greater the number of variations of the practice there are available for observation, the more secure the description of the constraint will become. The second way how

---

[10] It is however evident that, given a fixed set of constraints and a language-game defined with respect to these constraints, one cannot transform the language-game into another one which violates those constraints, any more than new rules of tennis can be inaugurated which go against the laws of physics.

[11] Wittgenstein was aware of this point. According to him, '[t]he *facts* of human natural history which cast light upon our problem are difficult to see because our discourse *passes them by* being busy with other matters. (Thus we say to someone 'Go to the shop and buy ...' — not 'Put your left foot in front of the right foot, etc. ... then put the money on the counter, etc. ...')' [his italics]. This is quoted in: G. P. Baker and P. M. S. Hacker, *Wittgenstein: Meaning and Understanding*, p. 586.

to make descriptions of constraints more reliable, a way which is certainly complementary to the first, is to engage in thought-experiments concerning the goal associated with the practice under investigation. One may ask: 'Is it conceivable to attain this particular goal if such and such a constraint were conventional rather than non-conventional?' If the answer is yes, the description of the constraint has to be modified or made more precise. This modification will result in a clearer distinction between conventional and non-conventional elements which were referred to together under the original, less precise description.

To recapitulate the line of argument developed in this section therefore: the aim was to arrive at a plausible account of non-conventional constraints. The starting point consisted in two qualifications to the preliminary ideas about constraints derived from Wittgenstein. The first concerned the idea of hinges; the second concerned what we can say about these hinges. The first qualification resulted in the conclusion that, if the non-conventional constraints on a given practice are to be described in terms of hinges, then they must be described as hinges that are context-independent or universal. This is a statement about practices in general. For the particular case of common practice, all constraints can be considered context-independent hinges because common practice was defined precisely as that practice which is universal because it lies very close to biological nature. The second qualification concerned the descriptions we can give of these constraints. If we take the description of the constraints to be well determinate and infallible then we would be taking the option of foundationalism according to which some statements, such as those about sensory perceptions, have a special role as foundations for all present and future theories. This option was avoided. It was shown that descriptions of non-conventional constraints are not well determined. These descriptions must take into consideration the entire combination of different aspects constituting the practice: perceptions, bodily movements, intentions, discourse, and so on. It was shown also that descriptions of non-conventional constraints are not infallible. Knowledge of constraints is always fallible and often tacit.

3. Objectivity

If this understanding of non-conventional constraint is accepted as plausible, how does it affect our understanding of objectivity? To answer this question is the aim of this final section. In the previous chapter, I presented a number of arguments to show that it is reasonable to claim that there is such a thing as common practice which is both universal and constrained exclusively by non-conventional constraints. In this chapter up to now I have been exploring the idea of constraints and I can now claim that the main conclusions of the previous chapter about common practice can still be held even though descriptions of the constraints involved are fallible. My proposal in this section is that these two characteristics, namely the characteristic of being universal and of being constrained exclusively by non-conventional constraints, are appropriate to determine the content of the notion of objectivity when applied to a practice. It will be helpful to start elaborating this proposal by situating it with respect to some aspects of the usual sense of objectivity.

According to a first straightforward understanding, objectivity is often attributed to things when they exist independently of any perception or representation of them. This is an ontological notion. It involves giving an account of the grounding of the thing's empirical reality. According to this view, all things should fall into two groups, some being clearly objective, like plants and rocks, and some being clearly subjective, like memories and dreams. A fictional concept, such as 'unicorn', would not have objective reality. The problematic nature of this view becomes evident when we realise that there may be quite a few things which tend to sit on the fence, like numbers, propositions, time and space. The status of such things can be the subject of lengthy debates.

A better move is to consider objectivity an attribute employed not to mark a distinction between types of entities but between types of cognitive achievement. What should be described as objective, as opposed to subjective, are judgements, or theories, rather than objects. Hence Kant describes objective validity as something not directly related to the external world but to the way judgements are made. If all inquirers agree on something, and they do this necessarily, then their judgement has objective validity:

> Objective validity therefore and necessary universality (for everybody) are equivalent notions, and though we do not know the object in itself, yet when we consider a judgement as universal, and also necessary, we understand it to have objective validity. By this judgement we cognise the object (though it remains unknown as it is in itself) by the universal and necessary connection of the perceptions given to us. As this is the case with all objects of sense, judgements of experience take their objective validity not from the immediate cognition of the object (which is impossible), but from the condition of universal validity in empirical judgements, which, as already said, never rests upon empirical, or, in short, sensuous conditions, but upon a pure concept of the understanding. The object always remains unknown in itself; but when by the concept of the understanding the connection of the representations of the object, which are given to

our sensibility, is determined as universally valid, the object is determined by this relation, and the judgement is objective.[12]

To see this clearly, it is useful to make some distinctions, which are probably the ones Kant himself presupposed in this and other parts of his work.[13] These distinctions concern the kind of vocabulary we use to classify the way an individual accepts the content of a proposition. If I accept X and do not have reasons acceptable to others in favour of X, then I am merely persuaded of X. If I accept X and I do have reasons acceptable to others in favour of X, then I am not only persuaded of X but also convinced of X. If I accept X and have reasons acceptable to others in favour of X, and moreover X is true, then I am certain of X. Hence, according to this view, conviction is more than persuasion, and certainty is more than conviction.[14] From the above quote, it seems plausible to suggest that by 'objective validity' Kant means intersubjective agreement in conviction where such agreement does not entail and is not entailed by the truth of the judgement about which persons share such agreement. By agreement here, I am suggesting that Kant can be taken to mean 'agreement for all possible rational inquirers'. Hence, when we say that a judgement is objectively valid, we are not saying that all rational inquirers actually endorse it. We are making a conditional assertion. We are saying that, if someone is a rational inquirer, then he or she would be convinced when endorsing it, as opposed to being merely persuaded.

This view does not go against the previous one involving entities. It is however broader. Thus a fictional concept, such as 'unicorn' would not have objective reality: it cannot be said to exist independently of the occurrence of representative states. However, it could very well function as a predicate in an objectively valid judgement, such as 'unicorns do not exist'.[15] By concentrating upon objectivity as an attribute of judgements, we are shifting our attention from talking about ontology to talking about cognitive achievement. Once this shift is accepted, we do

---

[12] I. Kant, *Prolegomena to any Future Metaphysics*, § 19.

[13] I am indebted to H. Allison, *Kant's Transcendental Idealism*; R. Meerbote, 'Kant's Use of the Notions "Objective Reality" and "Objective Validity"'.

[14] These distinctions have to do with reasoning about X and not with feeling X to be true. Hence, belief does not come into it. See the previous discussion in chapter three, §2.

[15] I am indebted to P. F. Strawson who identifies the important Kantian contrast between a very general conception of an object, which encompasses whatever can count as a particular instance of a general concept, and a 'weighty' sense, which applies only to what can be said to exist independently of the occurrence of representative states: *The Bounds of Sense*, pp. 73-74.

not ask: 'does object X exist?' but rather: 'under what conditions do we say that object X exists?' Hence we define what an object is in terms of conditions of a particular conception and not the other way round.[16]

The Kantian version of objectivity, although self-consistent, seems to have a vulnerable point. This is evident from the perspective of holism advocated up to now. Kant is giving an account of objectivity as an attribute of cognitive achievement. But such achievement cannot be conceived in the abstract. Cognitive achievement is about judgements, and for judgements we need statements. He presupposes that one is supplied with statements, and then proceeds to discuss the grounds on which one can attribute objective validity to the act of accepting them. However, judgements are here being taken to be about isolated propositions. Thoughts are considered more as isolated units than as linked to one another. Hence, if we are to employ the vocabulary introduced at the beginning of chapter four, we should say that one central concept he is using to give an account of objectivity, namely his idea of judgement, is not holistic. We realise that holism inevitably starts leaking in, as it were, when we recall that any judgement must be individuated by using words, and the meaning of these words may not be as secure and stable as Kant seems to suppose.

To understand better what is at issue here, allusion may be made to what has often been called objectivity of meaning.[17] The basic intuition behind this special kind of objectivity is that one cannot arrive at truth by changing the meaning of one's statements to suit the occasion. One cannot deliberate whether p, if the meaning of p is not fixed before the deliberation. In other words, the meaning of a statement is a real constraint. We are bound within a linguistic framework by contract, as it were. The usual understanding of this kind of objectivity is the following: a class of statements has objectivity of meaning if and only if the meaning of the statements is a real constraint to which those who utter or otherwise employ the statements are bound.

This account of objectivity of meaning is required by the Kantian treatment of objective validity. Only after one presupposes that the meaning of a statement is fixed does one attempt to decide whether a judgement concerning it has necessary universality or not. An examination of the way actual languages work will show however that objectivity of meaning is only an unreachable ideal. Objectivity of

---

16 Here I am following Kant: 'an object is that in the concept of which the manifold of given representations to an intuition is united ... it is the unity of consciousness that alone constitutes the relation of representations to an object, and therefore their objective validity.' *Critique of Pure Reason*, B 137.

17 C. Wright, *Realism, Meaning and Truth*, especially the Introduction, pp. 1-43.

meaning is not applicable to actual languages without some serious modifications, even though it can be applicable to formalised languages.

What shows this is the simple observation that the meaning of words sometimes changes over time. Some people, even perhaps single individuals by their literary reputation, do have a considerable role in changing the meaning of a particular word. In a very strong sense, we come from different backgrounds and we have to converge on our use of a given word. This is not very obvious when dealing with words like 'circle' and 'square'. But it becomes clearer when we are dealing with words like 'funny', or 'desirable'. This point was made in a slightly different way when discussing social holism in chapter three, §1. I mentioned there how our account of what the sense of a word is must take into account the fact that speakers often do not grasp the full meaning of the words they use. I argued that the best option is to hold that the sense of a word is a complex attribute that involves at least the three elements: first, the knowledge already possessed by the user, second, the knowledge of the experts associated with the relevant area of knowledge, and third, the criteria determining which people are to be acknowledged as experts.

An interesting and currently significant example is the application of the predicate 'intelligent'. This predicate has been employed to distinguish some mental activities, including those of human beings, from others. Can we apply the same attribute to computers? The usual, rather naive way of attempting to reply to this question is to assume that the meaning of 'intelligent' is somehow fixed by our previous use. All we have to do is to see whether computers, of this generation or of the next, will deserve the tag we usually wear ourselves, and when they will deserve it, they will get it — like some kind of medal. This view, however, is a misconception. The very existence of computers, of whatever generation, has an impact on what we mean by 'intelligent'. As artificial, inferential, creative and maybe even self-regularising algorithms are being developed, the notion of intelligence is probably changing as well. This change happens however in a way which allows, at any given moment of time, judgements to be made on whether a given activity, be it animal, human or artificial, corresponds to intelligence or not, in spite of the fact that the meaning of 'intelligent' itself depends on such activities. Such interaction is also observed in an example suggested by Crispin Wright:

[Objectivity of meaning] is the conception that the meaning of an expression stands to the unfolding tapestry of the way it is used in our linguistic practices as a person's character, according to a certain misconception of it, stands to his or her unfolding behaviour. The misconception would have it that character is, as it were, a finished design for a person's life which they usually act out, but which their behaviour may, at any particular stage, somehow betray. This has at least the virtue of explaining, but goes far beyond what is necessary to explain, the use we make of the idea that a person can act 'out of character'. But it is obvious enough that we also conceive of character as determined by behaviour: there are, we would like to say, *conceptual* limits to the extent and variety of ways in which a person can act out of character. A proper account of the relations between character and behaviour would have to display both how the nature of someone's character is a conceptual construct from what is said and done, and how it is nevertheless intelligible and fruitful to allow for the sort of contrast which we describe as 'acting out of character'.[18]

The upshot is that one cannot plausibly hold that our actual linguistic activity shows objectivity of meaning. The Kantian version of objective validity therefore needs some improvement. I suggest that such a refinement can be usefully carried out by making two important moves. The first one concerns going from a consideration of isolated judgements to a consideration of the entire web of belief that constitutes our knowledge. The second concerns the further inclusion of holism within the discussion by introducing the idea of an agglomeration of practices.

What justifies the first kind of refinement? Why should one feel obliged to make the move from an analysis of isolated beliefs to one dealing with the entire web of mutually dependent beliefs? One can perhaps try to avoid at all costs the modification of the notion of objective validity. This option involves returning to the Kantian program and attempting to consolidate his method of philosophising in terms of transcendental arguments in order to identify a priori principles. This option is not the one I will be taking here. The reason is that, even if it were effective — which in my opinion is doubtful — the option does not emphasise the cognitive and social holism I am committed to.[19] An option more in line with my approach is to try to save the notion of objective validity by seeing how it may be defined with respect to Quine's picture of a web of beliefs, some of which are at the centre and some at the periphery.

For Quine, it will be recalled, our entire knowledge can be seen as a sphere with observation sentences on the periphery and non-observation sentences in the interior, starting from comparatively ordinary sentences about material objects,

---

[18] C. Wright, 'Rule-Following, Meaning and Constructivism'.

[19] The point here is not that I am against all kinds of transcendental arguments. In fact, I resorted to such arguments in chapter six §1 to show that there is a core of criteria of rationality common to all inquirers, and hence a universal common practice. But here the question is not whether a common practice can be said to exist, but what a common practice consists of.

going on to more theoretical sentences to laws of physics and finally, at the centre, the laws of logic. If there is a disturbance at the periphery we must revise some observation sentences and also some other sentences on the interior. Our tendency, as Quine calls it, is to look for a comparatively minor adjustment at the periphery. Sentences nearer the centre are more resistant to revision. The laws of logic are the most resistant of all. Only extraordinary circumstances will convince us to abandon, say, the law of non-contradiction.

Now, a judgement is said to have objective validity if it is necessarily universal. The necessity aspect has no place in the picture suggested by Quine. In fact, on his view, any belief can be considered true if drastic enough changes are made within the system. Quine postulates a tendency of all inquirers to disturb the system as little as possible and to disturb it first at the periphery and then, only if needed, at the interior. The question one can ask is: why did Quine's picture place the laws of logic at the centre? What special attribute do these propositions have which qualifies them, rather than other propositions, to occupy the centre, and thus be most resistant to modifications? The answer that the previous paragraphs suggest is the following. The more a judgement is situated towards the interior, the more it is indispensable by the kind of inquirers we are. The judgements at the centre are presumably the most indispensable of all. Therefore, if we abstain from the Kantian a priori method of philosophising, it is these laws of logic situated at the core of the web of belief that seem to be the closest we can get to objectively valid judgements. So the answer to the question is the following. The judgements at the centre occupy that place because they are the most objectively valid ones.

If this first kind of refinement of our account of objectivity is accepted, then we notice that an aspect not directly included in the Kantian treatment of objective validity is hereby introduced, namely the aspect of degrees of objectivity. This aspect can now be accommodated. A plausible way how to understand it is as follows. A judgement A can be said to enjoy a higher degree of objectivity than a judgement B in the case when A is more indispensable than B for all possible inquirers like us.

This observation concludes the first step of refinement. Just as this first step involved the move towards a more holistic outlook expressed by the shift in emphasis from single judgements to a web of belief, so analogously the second step of refinement involves the move towards an even more comprehensive holism, this time expressed by the shift in emphasis from judgements in the abstract to practices. From what has been said about practice in chapter five, it seems natural now to go from the Quinean idea of our entire knowledge as being a web of belief to the corresponding idea of our human activity and discourse in its totality as made up of a complex agglomeration of practices. Again, just as in the Quinean web not all beliefs are on the same level, so also now we can envisage a kind of measure to differentiate between specific locations practices may occupy in the agglomeration of human activity. My reflections on the Kantian notion of objective validity have given good reasons to believe that this notion of objectivity can be useful in accounting for the fact that some judgements are situated at the core and some further outwards. The same notion of objectivity can also be useful in accounting for the fact that some practices can be considered at the centre of human activity and others further away. This can be done as follows.

A practice situated towards the centre would be one whose constraints are mainly non-conventional and one in which many human protagonists are engaged. Practices towards the periphery would be more local. Constraints acting on them are conventional and not many protagonists are engaged in them. In line with this view, given the conclusions reached in chapter six about common practice, and the those in the first section of this chapter, it is now reasonable to hold that the practice that lies at the centre of human activity and discourse, in the sense of being the most objective, is common practice. It will be recalled that common practice was defined as the one practice whose constraints are exclusively non-conventional and which is also the one engaged in by every human being, by the very fact of being a human being. Hence we can see that, according to the picture of human activity constituted of an agglomeration of practices, common practice, by definition, will be the most objective practice and it will hence be at the core.

Common practice will thus play the same role in the practice-view as the laws of logic did in the web-of-belief view. Laws of logic were considered as the most indispensable and least likely to be modified. Moreover, their location in the centre expresses the fact that they are presupposed by all laws and theories and statements as one moves out towards the periphery. Similarly, common practice can be viewed as consisting of rudimentary actions and behaviour patterns so closely linked to the kind of creatures we are and to the kind of world we live in that it is indispensable. Moreover it is engaged in, often indirectly, by all protagonists even when these protagonists are directly engaged in other practices whose constraints are conventional. Hence, just as anyone discussing whether some particular marks on the microscope slide are relevant to a given hypothesis is also indirectly endorsing the Law of Excluded Middle, so also anyone engaged in the practice of fashion design is also indirectly engaged in common practice. Another way to put this is that I am considering the purpose of survival as always present.

Two caveats should be made. First, although Quine's view of a web of belief, and the view suggested here of an agglomeration or cluster of practices are similar, especially in so far as a certain order is identified in terms of objectivity, there is at least one important difference. The difference concerns the role of the external world. For Quine, the periphery is where observation statements are situated. The outside world impinges on the most changeable statements and then modifications may or may not be needed at the deeper layers of our knowledge, consisting of theoretical statements and laws. For the picture involving a cluster of practices, the external world does not have a definite location. We cannot say that the practices towards the core are less impinged upon by the external world. In fact, any reference to the external world in this picture is kept to a minimum. The only, albeit significant, allusion to it is negative: it is acknowledged as relevant through the non-conventionality of some constraints. The second caveat concerns fallibilism. Not to fall into some kind of practice-foundationalism, one should recall that the description of constraints, including the description of the constraints on common practice, is always a fallible operation. Whether a constraint on a practice is conventional or not is something we can never be completely certain of.

Conclusion

I am now in a position to gather the consequences of these reflections into an overall conclusion to the whole chapter. It will be recalled that the aim was to determine how objectivity may be understood in terms of the constraints on a given practice. There were three sections. First, allusion was made to two relevant aspects from Wittgenstein. These were refined in the second section. The main valuable conclusion was that, for the particular case of common practice, all constraints can be considered context-independent hinges, in the sense used by Wittgenstein. In the second section, I showed moreover that descriptions of non-conventional constraints must take into consideration the entire combination of different aspects constituting the practice: perceptions, bodily movements, intentions, discourse, and so on. Such descriptions are certainly fallible.

The third and final section involved the application of these insights to our understanding of objectivity. The motivating idea was that, to determine the content of the notion of objectivity when applied to a practice, two attributes of the practice have a special role: the attribute of being universal and the attribute of being constrained exclusively by non-conventional constraints. To show this, two steps were necessary. A change was shown to be advantageous from an ontological version of objectivity to the Kantian one. Having taken this first step, we are not anymore attempting to mark a distinction between autonomous and dependent entities but between grades of cognitive achievement: if we endorse the Kantian notion, we focus our attention on judgements. The second step was the highlighting of the fact that endorsing the Kantian notion obliges us to endorse also an objectivity of meaning as regards the words we use to formulate these judgements. Moreover, it was shown that to assume there is such a thing as strict objectivity of meaning in actual languages is mistaken. The Kantian version of objective validity therefore needed some refinement. The suggestion was that such a refinement can be carried out by making two moves. The first one concerned going from a consideration of isolated judgements to a consideration of the entire web of accepted beliefs that constitutes our knowledge. The second concerned the further inclusion of holism by introducing the idea of an agglomeration of practices constituting our human activity. On this view, the practice that lies at the centre of human activity and discourse, in the sense of being the most objective, is common practice.

Chapter Eight

Scientific Objectivity and Realism

The original task I set myself in the first chapter was to give an account of scientific realism that does not need the correspondence theory of truth. The initial intuition was that this task can be accomplished by engaging in a detailed analysis of the role of holistic concepts in our understanding of science. Such concepts were shown to be unavoidable in some form or other. As regards this point, more precision was achieved by identifying a robust definition of holism. In chapter two, I was concerned with the cognitive aspect. Some definitions of holism already in use in the literature were first discussed and evaluated. Then the all-or-nothing problem was described. Two attempts at defending holism from this problem, namely molecularism and revisionism, were briefly presented, but it was shown that they both face formidable difficulties. Hence, the option was then taken to seek a justification only of a minimal version of holism which consists in the claim that it is inconceivable to have just one isolated thought. The justification of this minimal version made use of the claim that what allows a thought is a structure which depends on the biological makeup of the creature and also on a segment of its life: such a structure allows a certain disposition. According to the laws of nature as we know them these structures can only come about through slow growth. They must therefore accommodate more than one thought. This signifies that the mind, whose proper parts include thoughts, beliefs, and other propositional attitudes, is an aggregate with a very high degree of holism.

In parallel to this argument, chapter three was dedicated to the justification of social holism: the position according to which one cannot properly conceive of one isolated inquirer. In the first section it was shown that the sense of the words used by an inquirer to express his or her thoughts must bring in the entire community of inquirers in which those words are used successfully. This line of argument, although successful to a certain extent, was shown to be insufficiently general because it works only for certain words and not for others. A more general argument involved the investigation of the intentional attempt on the part of inquirers to shape their thoughts with a view to having them satisfy conditions of rationality. It was argued that, since an inquirer must follow rules which are knowable by others, we are obliged to hold that the inquirer has to check her rule-following by reference not only to her own responses but also to what others are inclined to do in the same domain. Hence, to properly conceive of one inquirer, other inquirers must be brought into the picture. I then argued that endorsing social holism does not lead to insurmountable problems which would emerge if an unrefined notion of group belief is used. The problems are avoided if for groups, instead of the notion of belief, one uses the notion of acceptance.

Chapter four consisted primarily in the analysis of two concepts which were called holistic because they guarantee that holism will not be neglected in our understanding of science: the notion of discourse, which is more related to cognitive holism, and the notion of practice, which is more related to social holism. The first section on discourse concluded with the claim that, even though discourses should not be considered to form a family in virtue of features common to all, one particular feature is of particular importance when dealing with science, namely the possibility of agreement and disagreement. The notion of cognitive command was brought in to elaborate on this attribute of discourses. The second section concerned the notion of practice. I defended this notion by showing that, when considering a given behaviour pattern which is usually explained by employing the notion of practice, something in common to all persons who show the pattern can indeed be legitimately spoken about even though each person had a unique route of acquiring that behaviour pattern. A multiplicity of different routes of acquisition does not necessarily mean different thing acquired. I also argued that essentialism should be avoided when dealing with practices. Hence there is no reason to replace the notion of practice by an exceedingly narrow notion of habit understood as referring to one particular causal chain of one individual's behaviour pattern. Habits are better understood as dispositions of people. Building on this, I arrived at the claim that practices can legitimately be taken to consist of actions with a certain purpose, actions manifesting habits.

In chapter five I applied these insights to the particular case of scientific practice. My aim was to offer some arguments in favour of the claim that science can be viewed, in some respects, as a single whole. An account of the purpose of scientific activity was first formulated. This allowed the claim that scientific practice is best seen as the set of actions grouped together (1) because they are geared towards the achievement of this unifying purpose, and (2) because they are performed in ways that involve habits or skills. The scientific discourse associated with this practice is constrained by conventional constraints and also by non-conventional constraints, the latter ones being of special significance for scientific objectivity. It was then shown that non-conventional constraints on scientific practice and discourse become manifest during the process of experimentation which involves ultimately the handling of objects in a laboratory just as we handle them in everyday life. The claim that there is an important link between science and the handling of objects in everyday life was defended against arguments that could arise from those who claim that there is a clear rupture between science and common sense.

So it became evident that the inclusion of holism in our understanding of science makes it imperative to explore the relation between scientific practice and common practice. Chapter six was dedicated to the investigation of the latter kind of practice. Common practice was understood as that practice in which common sense is embedded. Hence it includes handling things, moving around, and also the rudimentary kind of predicting that is needed to do so. I first presented a brief overview of some relevant points from the philosophical literature on common sense and on universal rationality. This enabled me to argue that the claim that there is a practice which is universal is plausible because the corresponding claim that a core of criteria of rationality is universal is plausible. I then refined the idea of common practice by alluding to Wittgenstein's notion of forms of life. The conclusion reached was that common practice is best seen as covering part of the area between two aspects of human living: the area between what is *culturally* natural, which is local, and what is *biologically* natural, which is universal.

In chapter seven the aim was to go deeper into the nature of constraints on practices to see how objectivity may be understood as an attribute of practices. It was shown that, to be consistent with the views expressed in the previous chapter, all the constraints on common practice should be considered not simply as Wittgensteinian hinges but as context-independent hinges. It was argued moreover that the process of giving clear descriptions of non-conventional constraints is revisable. Hence, to speak of objectivity using these terms, one has to make a number of moves. From an ontological version of objectivity one has to go to the Kantian version which marks a distinction between grades of cognitive achievement. Then, since this Kantian notion depends on the unwarranted assumption of objectivity of meaning, one has to refine it. My suggestion here was to include holism in our understanding of objectivity. This was done by alluding to the entire Quinean web of beliefs that constitutes our knowledge, and also by considering human activity more broadly as constituted of an agglomeration of practices. Hence, it was concluded that the practice that is the most objective and that lies at the centre of human activity and discourse is common practice. This is so because it is by definition universal and constrained exclusively by non-conventional constraints.

Now the next step to be made is to see what can be said about scientific practice if this practice is evaluated in terms of the notion of objectivity formulated above.

1. The Objectivity of Scientific Practice

It will be recalled that one of the merits of abandoning the ontological notion of objectivity and of endorsing the Kantian one is that we can now work with degrees of objectivity. When discussing accepted beliefs as part of a Quinean web, we can say that a judgement A enjoys a higher degree of objectivity than a judgement B in the case when A is more indispensable than B for all possible inquirers like us. When discussing practices, a practice P enjoys a higher degree of objectivity than a practice Q in the case when more of the constraints on P are non-conventional than those on Q. It was argued that, on these grounds, there is at least a minimal kind of order between practices within the entire agglomeration constituting human activity. The practice at the centre, which sets the standard of objectivity, is common practice.

So where does scientific practice stand? It can be shown that, within this framework, scientific practice turns out to be highly objective. Two points will be made to illustrate this. Firstly, one can see that scientific practice is universal. But universality here should not be understood in the sense that every inquirer is engaged in that particular practice. Scientific practice is universal in the sense that the methodological rules and general purpose which determine it ensure that the actions it involves are such that they *would* be done by any rational inquirer. Protagonists of scientific practice are only a subset of all existing inquirers. That is undeniable. Yet, if these protagonists act in a certain way, for example by doing X to come to know that Y, then the account of scientific practice arrived at in chapter five shows that doing X to come to know that Y is what any rational inquirer would do if he or she were in the same situation as these protagonists. This is the ideal towards which scientists aspire: it may not be reached at every moment of time. Disagreements about what to do in a particular context are common. Moreover, the experiments involved in confirming or falsifying theories sometimes cannot be repeated. The fact that Einstein's general theory of relativity received one of its early confirmations after the study of a total eclipse of the sun can be taken to show that the objectivity of scientific practice depends also on the assumption that the testimony of others is trustworthy. The plausible claim to make therefore is that scientific practice can be called universal in as much as this ideal of universality is realised. Secondly, I argued in chapter five §2 that many kinds of discourse are associated with scientific practice: some constrained, some free, some constrained conventionally and some constrained non-conventionally. It was also shown that the non-conventional constraints on scientific practice are constraints on the handling of objects during experimentation. Giving better and better descriptions of constraints is a fallible process, but, as time goes by, protagonists of scientific practice achieve more and more clarity as regards the distinction between conventional and non-conventional constraints. The claim that their actions are universal would thus gain more support. And therefore the practice gains in objectivity.

These two characteristics give strong indications that scientific practice can be considered highly objective. To consolidate this claim, more elaboration is in order. I will be exploring especially what is meant by saying that scientific practice achieves more and more clarity as regards the distinction between conventional and non-conventional constraints. My strategy will be the following. By saying that scientific practice achieves more and more clarity, one is suggesting that science is, in some way, self-adjusting. This suggestion, together with the fact that I am taking common practice, the one which is most objective, to be the practice in which common sense is embedded, recalls the philosophical position which is often called critical commonsensism. In fact, more light will be thrown on the understanding of science resulting from my overall analysis of holistic concepts if this position of critical common sense is alluded to. But some refinements to this position will be shown to be necessary.

I start by referring to the works of Charles Sanders Peirce, and I will concentrate mainly on two issues which are of special relevance to my approach.[1] The first issue concerns the relation between the position of critical common sense and what has been said regarding Wittgensteinian hinges. At one place in his writings, Peirce describes his position as follows:

[The defender of critical common-sense] opines that the indubitable beliefs refer to a somewhat primitive mode of life, and that, while they never become dubitable in so far as our mode of life remains that of somewhat primitive man, occasions of action arise in relation to which the original beliefs, if stretched to cover them, have no sufficient authority. In other words, we outgrow the applicability of instinct — not altogether, by any manner of means, but in our highest activities. [5.511]

---

[1] C. S. Peirce, 'Issues of Pragmatism'; 'Pragmaticism and Critical Common-Sensism'; 'Consequences of Critical Common-Sensism'. Subsequent reference to Peirce's writings will be according to the usual convention of volume number and paragraph number. Commentators consulted were: C. F. Delaney, *Science, Knowledge, and Mind*; C. Hookway, *Peirce*; C. J. Misak, *Truth and the End of Inquiry*. Peirce lacks clarity in his use of the word 'belief' because he does not make the distinction between belief and acceptance which I introduced in chapter three §2. When what he means is clear enough, I will retain his vocabulary.

Here we see Peirce claiming first that justification must come to a halt somewhere. The beliefs that provide the bedrock are beyond rational support, if having rational support is understood as having an articulatable reason. He also however envisages a feedback from activities other than our primitive interaction with the world. This latter aspect is what gives the extra critical element to the static version of common sense, according to which common-sense beliefs are completely immutable. Once we move beyond the sorts of cases involved in the normal circumstances of these beliefs, we have to be careful. Our common-sense beliefs help us cope with ordinary objects in everyday life. They provide us with useful tools in constructing physical theories which explain such ordinary objects. Peirce hopes that they provide us with useful tools also when science is dealing with objects that are too far away from any everyday experience. We consider electrons as objects in space, and we readily bring into our theorising some beliefs from our stock of common sense such as the belief that the motion of electrons is confined to three dimensions.[2] In thinking this way about electrons we are relying on common-sense beliefs in areas which are beyond those of their original setting. Peirce reminds us that we can only *hope* that these beliefs in their new setting will lead us to the truth.

It is important to remark as well that his later views mentioned that the bedrock can be considered a fixed set of beliefs, the same for all inquirers. There may be changes of an insignificant kind in this list of beliefs: the changes are so slow that for ordinary purposes they can be ignored.

> There are indubitable beliefs which vary a little and but a little under varying circumstances and in distant ages; that they partake of the nature of instincts, this word being taken in a broad sense; that they concern matters within the purview of the primitive man; that they are very vague indeed (such as, that fire burns) without being perfectly so.... [5.498]

It is evident that these views are not very different from the views expressed in the previous chapter regarding hinges, in the Wittgensteinian sense. The novelty here is that Peirce wants to accommodate a critical element within the idea of having a set of supporting hinges for one's knowledge and linguistic abilities. He elaborates this by observing that common-sense beliefs are very vague.

Vagueness, in fact, is the second issue I want to emphasise. According to Peirce, 'the Critical Common-sensist holds that all the veritably indubitable beliefs are *vague* — often in some directions highly so. Logicians have too much neglected the study of *vagueness*, not suspecting the important part it plays in mathematical

---

2 Peirce 5.445.

thought'.[3] Peirce gives a prominent role to vagueness because he is convinced that it is always present in language till the end of inquiry. His position is opposed to the view that vagueness should be eliminated at the beginning of inquiry for the reason that we cannot reason reliably until we have a precise language. Vagueness can be considered in either of two ways. In one sense, it can be taken to mean 'unspecific'. If a friend asks me for directions how to get to the railway station and I answer 'it is to the north', my answer would be unspecific because my friend needs more precise information. Similarly, the statement 'the chair weighs between 960g and 25461g' is under-specific because, although it involves the drawing of a sharp boundary, it draws it around a large area. In another sense, vagueness can be taken to mean the possession of borderline cases. The term 'clothing' is vague because a handkerchief can be definitely classified neither as clothing nor as non-clothing. Similarly, if I say 'I am of average height', I am drawing a blurred boundary around a small area thus introducing a lot of borderline cases. It is this second kind of unspecificity that corresponds to what Peirce, and many other philosophers, take as a satisfactory concept of vagueness. As regards the truth-conditions for such vague statements, the idea of a blurred boundary suggests that a useful contrast can be made between vague statements and general statements. A general statement is true if every way of determining it is true. Consider 'all humans are mortal'. Whatever shades of meaning of 'humans' and of 'mortal' we are allowed to pick out, we always end up with a true statement. Vague statements are different. Statements like 'I am of average height' or 'this week, a great event will happen' are true if *some* way of determining them results in a true statement. There are many criteria about how far away from the average one's height can be. There are many criteria about what kind of event is to be considered great. The two statements are vague because it is only according to *some* of these criteria that they turn out to be true. For Peirce, common-sense beliefs are vague and not general: of them, one can be certain because, when we say for example 'fire burns', we can always be sure that there is indeed some way of determining the meaning of the words in such a way that this common-sense belief turns out to be true.

In general, the more vague a proposition is, the fewer are the precise predictions that can be derived from it. One can transform a vague proposition into a more precise one. Then more falsifiable predictions can be made. Hence, the set of common-sense beliefs is considered to constrain scientific activity by supplying the vague propositions for science to render into a more precise form. If a scientific theory is falsified, this does not mean that common-sense beliefs are falsified as

---

[3] Peirce 5.505.

well. It means only that new precise formulations of the vague statements are needed. The relationship between a theoretical scientific understanding of the world and the set of beliefs constituting common sense is not one of the true to the false but one of the precise to the vague.

This position according to which scientific inquiry should be seen as related in this way to common-sense beliefs is described here exclusively by reference to the works of Peirce, but a similar position where scientific knowledge involves some strongly held core accepted beliefs to which other beliefs are held accountable has been endorsed also by many other philosophers of science as well, including Popper and Quine.[4] What the analysis of the holism discussed up to now indicates however is that this version of critical common sense is incomplete.

It is evident that Peirce is considering common sense as a set of accepted beliefs. One can indeed identify some of these beliefs, like the belief that fire burns. But what happens in everyday life, including all the interactions that are engaged in by humans with their surroundings, involves an innumerable variety of possibilities. It seems that, by taking common sense to be constituted of a set of accepted beliefs, Peirce and his followers need to presuppose that humans have in their possession an infinite number of beliefs stocked up and ready to be retrieved when needed. This leads to an uncomfortable position similar to the one mentioned in chapter three §2 as regards rule-following. There it was said that learning to follow a rule cannot certainly be equivalent to having an infinity of responses present within one's mind. Similarly, here it must be said that having common sense cannot certainly be equivalent to having an infinity of accepted beliefs present within one's mind — even if we concede that the beliefs are vague. Just as rule-following was convincingly accounted for by introducing an extrapolative tendency, so also here it seems plausible to consider common sense not in terms of a set of beliefs but in

---

[4] Popper takes up Neurath's analogy according to which inquirers are like sailors who must rebuild their ship in the open sea. Some parts of the ship are more difficult to repair than others given their pivotal structural roles. Quine's idea of a web of beliefs has already been discussed elsewhere. Although Peirce and Quine both emphasise the fact that some beliefs are the support of many others and are seldom, if ever, readjusted, the two philosophers are not employing the same kind of analogy. Quine's core beliefs are theoretical ones, like the law of excluded middle. For Quine, experience disturbs the system only at the edges, and leaves the interior underdetermined. Peirce's core common-sense beliefs are not beliefs constituting the *theoretical* substratum. They are vague beliefs about a certain subset of experience, namely everyday 'common' experience. This significant difference is unfortunately not apparent in the commentaries by Delaney and Hookway even though these authors highlight the parallelism between Peirce and Quine. Cf. C. F. Delaney, *Science, Knowledge, and Mind*, pp. 116-117; C. Hookway, *Peirce*, p. 230.

terms of habits which are manifested by certain accepted propositions. A normal person, having interacted with the world in a normal way, will acquire a disposition to form such and such accepted beliefs when confronted with such and such states of affairs. Peirce does not talk of common sense in this way. He does not highlight the interaction between, on the one hand, the beliefs he thinks constitute common sense, and, on the other hand, everyday actions and behaviour patterns. The closest he comes to admit this is when he emphasises how these beliefs are not theoretical and abstract but moulded in the experience of generations of human living. He explains, for example, that:

> those vague beliefs that appear to be indubitable have the same sort of basis as scientific results have. That is to say, they rest on experience — on the total everyday experience of many generations of multitudinous populations. Such experience is worthless for distinctively scientific purposes, because it does not make the minute distinctions with which science is chiefly concerned; nor does it relate to the recondite subjects of science, although all science, without being aware of it, virtually supposes the truth of the vague results of uncontrolled thought upon such experiences, cannot help doing so, and would have to shut up shop if she should manage to escape accepting them. No 'wisdom' could ever have discovered argon; yet within its proper sphere, which embraces objects of universal concern, the instinctive result of human experience ought to have so vastly more weight than any scientific result, that to make laboratory experiments to ascertain, for example, whether there be any uniformity in nature or no, would vie with adding a teaspoonful of saccharine to the ocean in order to sweeten it. [5.522]

It is evident therefore that the enlargement from the use of the notion of common sense to the use of the corresponding notion of common practice carried out in chapter six can profitably be carried out here as well.

My proposal, as should be expected by now, is to give an account of a critical mechanism at work within the interaction between common practice and scientific practice in the following way. Engaging in a practice, whatever that practice is, involves some awareness of the constraints acting on it. Since everyone engages in common practice, everyone has at least some tacit awareness of the constraints on common practice, which are non-conventional. This awareness can give rise to some vague descriptions of these constraints.

Hence, the constraints on common practice are constraints as well on scientific practice. On the theoretical level, this situation enables inquirers to formulate vague propositions for science to render into a more precise form. Rendering these descriptions more and more precise is a fallible process. The falsification of a scientific theory does not mean that we were mistaken in the way we engaged in common practice. It does not mean that we were mistaken in giving the constraints the vague descriptions that we do. It means only that we are called on to give new precise formulations so as to distinguish better between conventional and non-conventional elements.

One may object here that nothing is being added to the view already worked out by Peirce, because once some constraint on common practice is given a description, it is this description of the constraint that is transformed, via scientific practice, from a vague statement to a precise one. Hence, an objector may conclude that the account I have offered involving practices can be reduced without any loss to a purely Peircean operation on statements. The dimension of practice cancels out.

To respond to this objection, one must recall that, Peirce does not offer any explanation why common-sense beliefs are vague rather than precise. He just makes the observation and carries on with his investigation from there. What the approach developed so far supplies is precisely a convincing explanation of the Peircean observation that our common-sense beliefs start off as vague. From what has been said, one can see that the proposed explanation consists of the following points. (1) What constitutes a Peircean common-sense belief is a description of a constraint on common practice. (2) Protagonists of a practice in general, and therefore of common practice in particular, concentrate their attention on what they do and say — they are not *directly* concerned with the constraints that effectively limit what they do and say. (3) In common practice, this non-direct, subsidiary awareness of constraints results in descriptions which are not initially properly worked out in any detail but only expressed as sanctions, for example 'fire burns'. (4) The constituent words of the description of constraints allow for a very wide variety of determinations. In the example, 'fire' may be taken to mean 'whatever has a flame', or 'makes smoke', or 'gives light', or 'gives heat', or any combination of these features and possible others. Likewise, 'burns' may be taken to mean 'destroys everything', or 'causes severe pain', or 'destroys most things', or 'destroys these kinds of things but not those', and so on. Engaging in scientific practice enables protagonists to see which determination of the original vague description of the constraint is non-conventional. These four points show that the dimension of practice has a significant explanatory role supporting the Peircean account of common-sense beliefs.

So one can safely conclude that my analysis has resulted in a reasonably satisfactory position which includes a process whereby scientific practice achieves more and more clarity as regards the distinction between conventional and non-conventional aspects. Not emphasising holism in our understanding of science can lead to the Peircean position according to which science involves inquirers engaging in critical common sense. Emphasising holism can lead to the corresponding position according to which science involves inquirers engaging in critical common practice.

2. The Original Task

The main question in the first chapter concerning the defensibility of scientific realism can now be re-evaluated. The original aim had been to show how holism enables realists to successfully defend their position. This position was characterised by two claims: first that the external world exists, and second that we are capable of coming to know something about it through scientific inquiry. Endorsing the correspondence theory of truth to enable the holding of both claims was shown to be a questionable move. What realists need is an explanation of how scientific inquiry can result in judgements that are true or false not because of the inquirer's standpoint but because of something neutral to all. If scientists have enough grounds for objectivity, then scientific realism would be plausible. Investigation into the role of the holistic concepts of discourse and practice has given enough grounds to hold that to consider scientific practice highly objective is indeed plausible. Hence realism is plausible to the same extent.

The formulation of realism, however, has to be modified. Discussions concerning scientific realism often involve investigations about how a specific statement in a theory can tell us something about the external world. Now, through the inclusion of holism both cognitive and social, the outlook is significantly different. Having emphasised holism in our understanding of science, we can now see that realism is no longer a philosophical position that deals with whether a statement within a theory refers or not. It is no longer a position that deals with whether a theory approximately says what the external world is actually like. When viewed from the perspective of holism, realism becomes a philosophical position that deals with the entire network of interdependent theories and activities constituting scientific practice.

To appreciate the merits of this new version of realism, it is useful to mention that it does not inherit the problems of the correspondence theory of truth mentioned in chapter one. It will be recalled that there were two major problems. The first concerned the necessity of representation. If correspondence truth is understood as being some kind of comparison between particular judgements on one hand and pieces of the world on the other, then one seems to be implying that pieces of the world can be apprehended and compared without being represented: which is a very problematic implication if not an outright impossibility. This problem does not arise anymore because we are now committing ourselves neither to particular entities, in need of representing, nor to an ontological understanding of objectivity attributable to entities. We are not committing ourselves to this even though when speaking of non-conventional constraints we are in fact implying that the external world, as that phrase is often understood, exists. The second major problem discussed in chapter one concerned the apparently inevitable question-begging moves incorporated within accounts of correspondence-truth. It was shown how accounts that were successful in avoiding the Scylla of the representation problem, could not avoid the Charybdis of question-begging: they included the very notion of correspondence they were meant to account for in the first place. This kind of circular argument does not make an appearance within the present understanding of realism because correspondence is not included.

The overall analysis of the role of the two main holistic concepts of discourse and practice has therefore resulted in a particular understanding of science. This understanding adds considerable plausibility to the claim that scientific realists can still argue their case even if they do not presuppose the validity of the correspondence theory of truth.

Bibliography

Allison, H., *Kant's Transcendental Idealism*, New Haven: Yale University Press, 1983.

Anscombe, G. E. M., 'The First Person', in: S. Guttenplan, (ed.), *Mind and Language*, pp. 45-65; reprinted in: Q. Cassam, (ed.), *Self-Knowledge*, Oxford: Oxford University Press, 1994, pp. 140-159.

Aronson, J., *A Realist Philosophy of Science*, London: Macmillan, 1984.

Atran, S., *Cognitive Foundations of Natural History: towards an Anthropology of Science*, Cambridge: Cambridge University Press, 1990.

Austin, J. L., 'Truth', *Proceedings of the Aristotelian Society*, suppl. vol. 24 (1950); reprinted in: G. Pitcher, (ed.), *Truth*.

Baker, G. P., and Hacker P. M. S., *Wittgenstein: Meaning and Understanding. An analytical Commentary on the Philosophical Investigations, volume 1*, Oxford: Blackwell, 1980.

— *Wittgenstein: Rules, Grammar and Necessity. An analytical Commentary on the Philosophical Investigations, volume 2*, Oxford: Blackwell, 1985.

Bambrough, R., 'Universals and Family Resemblances', *Proceedings of the Aristotelian Society* 61 (1960-61), pp. 207-222.

Belinfante, F. J., *A Survey of Hidden Variables Theories*, Oxford: Pergamon Press, 1973.

Bilgrami, A., *Belief and Meaning, the Unity and Locality of Mental Content*, Oxford: Blackwell, 1992.

Blackburn, S., 'The Individual strikes back', *Synthese* 58 (1984), pp. 281-301.

Boghossian, P., 'The Rule-Following Considerations', *Mind* 98 (1989), pp. 504-550.

Boyd, R., 'What Realism implies and what it does not', *Dialectica* 43 (1989), pp. 5-29.

— 'The current Status of Scientific Realism', in: J. Leplin, (ed.), *Scientific Realism*, 1984, pp. 41-82.

Buchwald, J., 'Design for Experimenting', in: P. Horwich, (ed.), *World Changes*, Cambridge Mass.: MIT Press, 1993, pp. 169-206.

Carnap, R., *The Logical Syntax of Language*, London: Routledge and Kegan Paul, 1937.

Cartwright, N., 'Can Wholism reconcile the Inaccuracy of Theory with the Accuracy of Prediction?', *Synthese* 89 (1991), pp. 3-13.
— 'Fundamentalism vs. the Patchwork of Laws', *Proceedings of the Aristotelian Society* 94 (1994), pp. 279-292.
Caruana, L., 'John von Neumann's "Impossibility Proof" in a Historical Perspective', *Physis* 32 (1995), pp. 109-124.
— 'Beyond the Internal Realist's Conceptual Scheme', *Metaphilosophy* 27 (1996), pp. 296-301.
Cohen, L. J., *An Essay on Belief and Acceptance*, Oxford: Clarendon Press, 1992.
Craig, E. J., 'The Problem of Necessary Truth', in: S. Blackburn, (ed.), *Meaning, Reference and Necessity*, Cambridge: Cambridge University Press, 1975, pp. 1-31.
— *Knowledge and the State of Nature*, Oxford: Clarendon Press, 1990.
Davidson, D., 'On the very Idea of a Conceptual Scheme', *Proceedings and Addresses of the American Philosophical Association* 57 (1974), pp. 5-20; reprinted in: *Inquiries into Truth and Interpretation*, 1984, pp. 183-198.
— 'Mental Events', in: *Essays on Actions and Events*, Oxford: Clarendon Press, 1980, pp. 207-227.
— 'Semantics for Natural Languages', in: *Inquiries into Truth and Interpretation*, 1984, pp. 55-64.
— 'Theories of Meaning and Learnable Languages', in: *Inquiries into Truth and Interpretation*, 1984, pp. 3-15.
— *Inquiries into Truth and Interpretation*, Oxford: Clarendon Press, 1984.
— 'A Nice Derangement of Epitaphs', in: E. Lepore (ed.), *Truth and Interpretation: Perspectives on the Philosophy of Donald Davidson*, Oxford: Blackwell, 1986, pp. 434-446.
Delaney, C. F., *Science, Knowledge, and Mind*, Notre Dame, Ind.: University of Notre Dame, 1993.
Duhem, P., *The Aim and Structure of Physical Theory*, trans. P. P. Wiener, Princeton: Princeton University Press, 1954.
Dummett, M., 'What is a Theory of Meaning? (I)', in: S. Guttenplan (ed.), *Mind and Language*, 1975, pp. 97-122.
— 'What is a Theory of Meaning? (II)', in: G. Evans and J. McDowell, (eds.), *Truth and Meaning*, Oxford: Clarendon Press, 1976, pp. 67-137.
— 'The Philosophical Basis of Intuitionistic Logic', in: *Truth and Other Enigmas*, 1978, pp. 215-247.
— 'The Social Character of Meaning', in: *Truth and other Enigmas*, 1978, pp. 420-430.

Dummett, M., 'The Justification of Deduction', in: *Truth and Other Enigmas*, 1978, pp. 290-318.

— *Truth and Other Enigmas*, London: Duckworth, 1978.

Esfeld, M., 'Holism and Analytic Philosophy', *Mind* 107 (1998), pp. 365-380.

Evnine, S., *Donald Davidson*, Oxford: Polity Press, 1991.

Fodor, J., and Lepore, E., *Holism, a Shopper's Guide*, Oxford: Blackwell, 1992.

Garfinkel, A., *Forms of Explanation*, New Haven: Yale University Press, 1981.

Gilbert, M., 'Modelling Collective Belief', *Synthese* 73 (1987), pp. 185-204.

— *Social Facts*, London: Routledge, 1989.

Gooding, D., *Experiment and the Making of Meaning*, Dordrecht: Kluwer Academic Press, 1990.

Greenwood, J. D., 'Two Dogmas of Neo-Empiricism: the "Theory-Informity" of Observation and the Quine-Duhem Thesis', *Philosophy of Science* 57 (1990), pp. 553-574.

Grice, H. P., and Strawson, P. F., 'In Defence of a Dogma', *Philosophical Review* 65 (1956), pp. 141-158.

Grünbaum, A., 'The Falsifiability of Theories: total or partial? A Contemporary Evaluation of the Duhem-Quine Thesis', *Synthese* 14 (1962), pp. 17-34.

Guttenplan, S., (ed.), *Mind and Language: Wolfson College Lectures 1974*, Oxford: Oxford University Press, 1975.

Hacker, P. M. S., *Insight and Illusion*, Oxford: Clarendon Press, 1986.

Haldane, J., and Wright, C., (eds.), *Reality, Representation, and Projection*, Oxford: Oxford University Press, 1993.

Harré, R., *Great Scientific Experiments*, Oxford: Oxford University Press, 1981.

Harrison, R., *On what there must be*, Oxford: Clarendon Press, 1974.

Hartshorne, C., Weiss, P., and Burks, A., (eds.), *Collected Papers of Charles Sanders Peirce*, Cambridge Mass.: Harvard University Press, 1931-1958.

Heal, J., *Fact and Meaning*, Oxford: Basil Blackwell, 1989.

— 'Semantic Holism: still a Good Buy', *Proceedings of the Aristotelian Society* 94 (1994), pp. 325-339.

Hellmann, G., 'Realist Principles', *Philosophy of Science* 50 (1983), pp. 227-249.

Hertzberg, L., 'On the Factual Dependence of the Language-Game', *Acta Philosophica Fennica* 28 (1976), pp. 126-153; reprinted in: J. Canfield, (ed.), *Knowing, Naming, Certainty and Idealism. (The Philosophy of Wittgenstein, Volume 8)*, New York: Garland Publishing Company, 1986, pp. 290-317.

Hollis, M., and Lukes, S., (eds.), *Rationality and Relativism*, Oxford: Blackwell, 1982.

Holtzman, S., and Leich, C., (eds.), *Wittgenstein To Follow a Rule*, London: Routledge and Kegan Paul, 1981.

Hookway, C., *Peirce*, London: Routledge and Kegan Paul, 1985.

— *Quine*, Oxford: Polity Press, 1988.

Horwich, P., 'Three Forms of Realism', *Synthese* 51 (1982), pp. 181-201.

Hunter, J. F. M., '"Forms of Life" in Wittgenstein's Philosophical Investigations', in: E. D. Klemke, (ed.), *Essays on Wittgenstein*, Illinois: University of Illinois Press, 1979, pp. 273-297.

Jackson, F., and Pettit, P., 'Program Explanation: a General Perspective', *Analysis* 50 (1990), pp. 107-117.

Jammer, M., *Concepts of Mass*, Cambridge Mass.: Harvard University Press, 1961.

— *The Philosophy of Quantum Mechanics*, New York: Wiley-Interscience Publications, 1974.

Jardine, N., *The Fortunes of Inquiry*, Oxford: Clarendon Press, 1986.

— *The Scenes of Inquiry*, Oxford: Clarendon Press, 1991.

Joravsky, D., *The Lysenko Affair*, Cambridge Mass.: Harvard University Press, 1970.

Kant, I., *Prolegomena to any Future Metaphysics*, trans. L. W. Beck, Indianapolis: Bobbs-Merrill, 1950.

— *Critique of Pure Reason*, trans. N. Kemp Smith, London: Macmillan, 1964.

Klee, R., 'In Defence of the Quine-Duhem Thesis: a Reply to Greenwood', *Philosophy of Science* 59 (1992), pp. 487-491.

Kripke, S., *Wittgenstein on Rules and Private Language*, Oxford: Basil Blackwell, 1982.

Kuhn, T., *The Structure of Scientific Revolutions*, Chicago: University of Chicago Press, 2nd ed., 1970.

Lakatos, I., 'The Problem of appraising Scientific Theories: three approaches', in: *Mathematics, Science and Epistemology*, Philosophical Papers vol. 2, J. Worrall and G. Currie, (eds.), Cambridge: Cambridge University Press, 1978, pp. 107-120.

— 'Popper on Demarcation and Induction', in: *The Methodology of Scientific Research Programmes*, 1978, pp. 139-167.

— *The Methodology of Scientific Research Programmes*, Philosophical Papers vol. 1, J. Worrall and G. Currie, (eds.), Cambridge: Cambridge University Press, 1978.

Lakatos, I., and Zahar, E., 'Why did Copernicus's Research Programme supersede Ptolemy's?', in: R. Westman, (ed.), *The Copernican Achievement*, Los Angeles:

University of California Press, 1976, pp. 354-383; republished in: *The Methodology of Scientific Research Programmes*, 1978, pp. 168-192.

Laudan, L., *Progress and its Problems*, London: Routledge, 1977.

— 'A Confutation of Convergent Realism', in: J. Leplin, (ed.), *Scientific Realism*, 1984, pp. 218-249.

Lawrence, C., and Weisz, G., (eds.), *Greater than the Parts: Holism in Biomedicine 1920-1950*, New York, Oxford: Oxford University Press, 1998.

Lear, J., 'The Disappearing "We"', *Aristotelian Society Supplementary Volume* LVIII (1984), pp. 219-242.

— 'Transcendental Anthropology', in: P. Pettit and J. McDowell, (eds.), *Subject, Thought, and Context*, 1986, pp. 267-298.

Leplin, J., (ed.), *Scientific Realism*, Berkeley: University of California Press, 1984.

Lerner, D., (ed.), *Parts and Wholes*, New York: The Free Press of Glencoe, 1963.

Lewis, D., *Convention*, Cambridge Mass.: Harvard University Press, 1969.

— 'Radical Interpretation', *Synthese* 23 (1974), pp. 331-344; reprinted in: *Philosophical Papers*, vol. 1, New York: Oxford University Press, 1983.

Lipton, P., 'Contrastive Explanation', in: D-H. Ruben, (ed.), *Explanation*, Oxford: Oxford University Press, 1993.

Lugg, A., 'Disagreement in Science', *Zeitschrift für allgemeine Wissenschaftstheorie* IX/2 (1978), pp. 276-292.

MacIntyre, A., *After Virtue*, London: Duckworth, 2nd edition, 1985.

Mauss, M., 'Body Techniques', in: *Sociology and Psychology: Essays*, trans. B. Brewster, London: Routledge and Kegan Paul, 1979, pp. 97-135.

Meerbote, R., 'Kant's Use of the Notions "Objective Reality" and "Objective Validity"', *Kant-Studien* 63 (1972), pp. 51-58.

Mellor, D. H., 'In Defence of Dispositions', *Philosophical Review* 83(2) (1974), pp. 157-181.

— 'The Reduction of Society', *Philosophy* 57 (1982), pp. 51-75.

Misak, C. J., *Truth and the End of Inquiry*, Oxford: Clarendon Press, 1991.

Moore, G. E., 'Proof of an External World', in: *Philosophical Papers*, London: Allen and Unwin, 1959.

Moulines, C. U., 'The Ways of Holism', *Noûs* 20 (1986), pp. 313-330.

Nersessian, N. J., *Faraday to Einstein: Constructing Meaning in Scientific Theories*, Dordrecht: Martinus Nijhoff Publishers, 1984.

Newton-Smith, W. H., *The Rationality of Science*, London: Routledge and Kegan Paul, 1981.

— 'Modest Realism', *Philosophy of Science*, vol. 2 (1988), pp. 179-189.

O'Connor, D. J., *The Correspondence Theory of Truth*, London: Hutchinson, 1975.

Peirce, C. S., 'Issues of Pragmaticism', in: *Collected Papers*, C. Hartshorne, P. Weiss and A. Burks, (eds.), vol. 5, 1931-1958, pp. 293- 316.

—'Pragmaticism and Critical Common-Sensism', in: *Collected Papers*, C. Hartshorne, P. Weiss and A. Burks, (eds.), vol. 5, 1931-1958, pp. 346-350.

—'Consequences of Critical Common-Sensism', in: *Collected Papers*, C. Hartshorne, P. Weiss and A. Burks, (eds.), vol. 5, 1931-1958, pp. 351-375.

— 'Hume on Miracles', in: *Collected Papers*, C. Hartshorne, P. Weiss, and A. Burks, (eds.), vol. 6, 1931-1958, pp. 356-369.

Pettit, P., 'The Reality of Rule-Following', *Mind* 99 (1990), pp. 1-21.

— *The Common Mind*, Oxford: Oxford University Press, 1993.

Pettit, P., and McDowell, J., (eds.), *Subject, Thought and Context*, Oxford: Oxford University Press, 1986.

Phillips, D. C., *Holistic Thought in Social Science*, Stanford CA.: Stanford University Press, 1976.

Phillips, D. L., *Wittgenstein and Scientific Knowledge*, London: Macmillan, 1977.

Pitcher, G., (ed.), *Truth*, Englewood Cliffs, New Jersey: Prentice Hall, 1964.

Polanyi, M., *Personal Knowledge*, London: Routledge and Kegan Paul, 1958.

Popper, K., *Conjectures and Refutations: The Growth of Scientific Knowledge*, London: Routledge and Kegan Paul, 1972.

— *Objective Knowledge*, Oxford: Clarendon Press, 1979.

Putnam, H., *Mind, Language and Reality*, Cambridge: Cambridge University Press, 1975.

— *Meaning and the Moral Sciences*, Cambridge: Cambridge University Press, 1978.

— *Reason, Truth and History*, Cambridge: Cambridge University Press, 1981.

— *Realism and Reason*, Cambridge: Cambridge University Press, 1983.

Quigg, C., 'The SSC: Scientific Motivation and Technical Progress', in: K. Eggert, H. Faissner, and E. Radermacher, (eds.), *Sixth Topical Workshop on Proton-Antiproton Collider Physics 30th June - 4th July, 1986*, Singapore: World Scientific, 1987, pp. 736-761.

Quine, W. V. O., 'Two Dogmas of Empiricism', in: *From a Logical Point of View*, Cambridge Mass.: Harvard University Press, 1953, pp. 20-46.

— *Word and Object*, Cambridge Mass.: MIT Press, 1960.

— *Theories and Things*, Cambridge Mass.: Harvard University Press, 1981.

— *Pursuit of Truth*, Cambridge Mass.: Harvard University Press, 1990.

Ritson, D., 'Demise of the Texas Supercollider', *Nature* 366 December, 1993, pp. 607-610.

Rosenberg, J., *Linguistic Representation*, Dordrecht: D. Reidel Publishing Company, 1974.

— *One World and our Knowledge of it*, Dordrecht: D. Reidel Publishing Company, 1980.

Ruben, D-H., *The Metaphysics of the Social World*, London: Routledge and Kegan Paul, 1985.

Ryle, G., 'Knowing how and Knowing that', *Proceedings of the Aristotelian Society* 46 (1945), pp. 1-16.

Sanders, A. F., *Michael Polanyi's Post-critical Epistemology*, Amsterdam: Rodopi, 1988, pp. 138-145.

Sellars, W., *Science, Perception and Reality*, London: Routledge and Kegan Paul, 1963.

Shapin, S., and Schaffer, S., *Leviathan and the Air Pump: Hobbes, Boyle and the Experimental Life*, Princeton: Princeton University Press, 1985.

Stevenson, L., and Byerly, H., *The Many Faces of Science*, Oxford: Westview Press, 1995.

Strawson, P. F., 'Truth', *Proceedings of the Aristotelian Society*, suppl. vol. 24 (1950); reprinted in: G. Pitcher, (ed.), *Truth*.

— *The Bounds of Sense, an Essay on Kant's Critique of Pure Reason*, London: Methuen, 1966.

Stroud, B., *The Significance of Philosophical Scepticism*, Oxford: Clarendon Press, 1984.

Tennant, N., 'Holism, Molecularity and Truth', in: B. M. Taylor, (ed.), *Michael Dummett, contributions to philosophy*, Dordrecht: Martinus Nijhoff Publishers, 1987, pp. 31-58.

Thornton, T., *Judgement, Objectivity and Practice*, unpublished Ph.D. thesis, University of Cambridge, 1994.

Turner, S., *The Social Theory of Practices: Tradition, Tacit Knowledge and Presuppositions*, Oxford: Polity Press, 1994.

van Fraassen, B. C., *The Scientific Image*, Oxford: Clarendon Press, 1980.

Varney, R. N., 'Kinetic Theory', in: R. G. Lerner and G. L. Trigg, (eds.), *Encyclopedia of Physics*, Cambridge: VCH Publishers, 2nd edition, 1990, pp. 601-605.

Williams, B., 'Wittgenstein and Idealism', in: *Moral Luck*, Cambridge: Cambridge University Press, 1982.

Williamson, T., *Vagueness*, London: Routledge, 1994.

Wilson, B. R., (ed.), *Rationality*, Oxford: Blackwell, 1970.

Wittgenstein, L., *Philosophical Investigations*, trans. G. E. M. Anscombe, Oxford: Blackwell, 1953.

— *Remarks on the Foundations of Mathematics*, trans. G. E. M. Anscombe, Oxford: Blackwell, 1956.

— *Tractatus Logico-Philosophicus*, trans. D. F. Pears and B. F. McGuinness, London: Routledge and Kegan Paul, 1961.

— *Zettel*, trans. G. E. M. Anscombe, Oxford: Blackwell, 1967.

— *On Certainty*, trans. D. Paul and G. E. M. Anscombe, Oxford: Blackwell, 1969.

— *Philosophical Grammar*, trans. A. Kenny, Oxford: Blackwell, 1974.

Woolgar, S., *Science: the very Idea*, Chichester: Ellis Horwood, 1988.

Wright, C., (ed.), *Special Issue on Rule-Following, Synthese* 58 (1984).

— 'A Cogent Argument against Private Language?' in: P. Pettit and J. McDowell, (eds.), *Subject, Thought and Context*, 1986, pp. 209-266.

— *Realism, Meaning and Truth,* Oxford: Blackwell, 1986.

— 'Rule-Following, Meaning and Constructivism', in: C. Travis, (ed.), *Meaning and Interpretation*, Oxford: Basil Blackwell, 1986, pp. 273-274.

— 'Realism, Antirealism, Irrealism, Quasi-Realism', in: P. A. French, T. E. Uehling and H. K. Wettstein, (eds.), *Midwest Studies in Philosophy* vol. 12, Minneapolis: University of Minnesota Press, 1988, pp. 25-49.

— *Truth and Objectivity*, Cambridge Mass.: Harvard University Press, 1992.

Yoshida, R. M., *Reduction in the Physical Sciences*, Halifax, Nova Scotia: Dalhousie University Press, 1977.

Index

Printed and bound by CPI Group (UK) Ltd, Croydon, CR0 4YY

22/10/2024

01777620-0011